Michael Faraday

The Great Scientist Who Never Went to College

(Theories about Electromagnetism and Electrochemistry)

Frank Snyder

Published By **Darby Connor**

Frank Snyder

Michael Faraday: The Great Scientist Who Never Went to College (Theories about Electromagnetism and Electrochemistry)

ISBN   978-1-998901-75-3

Legal & Disclaimer

The information contained in this book is not designed to replace or take the place of any form of medicine or professional medical advice. The information in this book has been provided for educational & entertainment purposes only.

The information contained in this book has been compiled from sources deemed reliable, and it is accurate to the best of the Author's knowledge; however, the Author cannot guarantee its accuracy and validity and cannot be held liable for any errors or omissions. Changes are periodically made to this book. You must consult your doctor or get professional medical advice before using any of the suggested remedies, techniques, or information in this book.

Table of Contents

Chapter 1: Who Was Michael Faraday? ... 1

Chapter 2: His Personal Life and Early Years 5

Chapter 3: Electrochemistry 13

Chapter 4: Electromagnetism 34

Chapter 5: Civil Service and the Royal Organization 43

An Unconventional Path 50

Inventing and Educating 101

Further Creations and Legacies 122

Chapter 1: Who Was Michael Faraday?

Michael Faraday FRS was an English researcher who made significant contribution to electrochemistry as well as electromagnetism. The ideas that underlie electro-magnetic induced diamagnetism, electrolysis, and diamagnetism were among his most important discoveries.

Regardless of the fact that he did not have an formal training, Faraday was just one of the greatest scholars in the world.

Faraday established the framework for the concept of electromagnetic field in physics by his research on the electromagnetic field surrounding the conductor that is with direct current. Faraday also discovered that magnetic fields could influence beams, and that the two phenomena were inextricably linked. 3 He also developed the electrolysis guidelines and the theories of electro-magnetic induction and diamagnetism. Electro-magnetic rotary gadgets that he invented made him a pioneer in the field of electrical motor innovation and it was due to his efforts that electricity proved to be advantageous in the field of the field of innovation.

Faraday was an chemist who discovered the chemical benzene, studied chlorine clathrate the hydrate, and developed an early version on the Bunsen burner as well as the oxidation number system and promoted terms such as "anode," "cathode," "electrode," and "ion." Faraday was eventually the Royal Organization's very first and principal Fullerian professor of Chemistry and Chemistry, a position that he held for all of his existence.

Faraday was a brilliant experimenter, who expressed his ideas in a simple and straightforward language. Unfortunately his math abilities only covered the simplest algebra, and did not extend beyond trigonometry. James Clerk Maxwell manufactured the work of Faraday and other scientists into formulas which are now recognized as the basis of all current electro-magnetic theories. Maxwell claimed that Faraday's use in the form of line of force proved that the man is "in reality a mathematician of a really high order-- one from whom future mathematicians could get beneficial and fertile ways." Faraday is the SI capacitance system named after him.

On the wall of his research, Albert Einstein hung a photograph of Faraday alongside pictures of Arthur Schopenhauer and James Clerk Maxwell.

6 "There is no respect too big to pay to the memory of Faraday, one of the best clinical originators of perpetuity," said the physicist Ernest Rutherford, "when we consider the scale and variety of his

discoveries and their impact on the development of science and market."

Chapter 2: His Personal Life and Early Years

Michael Faraday was born in Newington Butts, Surrey, on the 22nd of September, 1791. (which has now become part of the London District of Southwark). He was born in a poor family. His father, James, was a Christian who belonged to the Glasite sect. In the winter of 1790 James Faraday moved his marriage couple and two children to Outhgill, Westmorland, where the former apprentice had worked to the town's blacksmith. In the autumn of 1790, Michael was born. Michael Faraday, the 3rd of four kids, was required to be educated after receiving only a basic education.

He began working as an apprentice with George Riebau, a regional bookbinder and bookseller from Blandford Street, when he was 14 years old.

Faraday spent a great deal of time reading writing during his seven-year apprenticeship including Isaac Watts' "The Improvement of the Mind", which he was adamantly embracing the concepts and guidelines in. Faraday also became attracted by the science of physics, especially electricity. The book by Jane Marcet Discussions on Chemistry was specifically stimulating to Faraday.

Let's look at what else the man did. In this complete guide, you'll discover more.

Faraday attended lectures given by the well-known English chemical scientist Humphry Davy of the Royal Organization and the Royal Society, and John Tatum the founder of the City Philosophical Society, in 1812, when he was just twenty years old, and was close to the end of his training. William Dance, one of the founders of the Royal Philharmonic Society was the one who gave Faraday lots of tickets to these talks. Faraday also presented Davy the 300-page book that was built on notes he took during these lectures. Davy responded quickly, enthusiastically and kindly. After Davy's vision was damaged in an accident involving nitrogen trichloride during the year 1813, he decided to hire Faraday for his position as assistant. On the same time that an assistant at the Royal Organization John Payne, was fired Sir Humphry was advised to locate an inheritor. Because of this the 1st of March, 1813, he named Faraday as a Chemical Assistant for the Royal Organization. Davy quickly transferred Faraday's duties in the form of samples of nitrogen trichloride and both were injured by the extremely spongy material that went up in flames.

On June 12th, 1821, Faraday wed Sarah Barnard (1800-- 1879).

They were married in the Sandemanian church with their families. He declared his belief to the Sandemanian church a month later, after their wedding. They didn't have children.

Faraday was a passionate Christian who was a member of the Sandemanian sect of the Church of Scotland. He was deacon as well as a senior in the house of meetings in which he was raised for two terms following his wedding. Paul's Street in the Barbican was the site which his parish was situated in. The meeting house was relocated into Barnsbury Grove, Islington, in 1862 and it was in this location where Faraday served for the final two years of his second period as a senior prior to resigning.

According biographical biographers "a deep sense of God and nature's unity bathed Faraday's life and work."

Faraday was honored with with an Honorary Medical expert in the field of Civil Law degree by the University of Oxford in the month of June 1832. Through his entire life he was awarded an honorary knighthood to recognize his contributions to science. which he declined due to spiritual grounds, believing that accumulating wealth and seeking for a worldly gain was a blatant violation of the commandments of God and he decided to remain "easy Mr Faraday to the end." He was elected a member of the Royal Society in the year 1824. However, he was unable to become President on two occasions. In 1833, he received the honor of being designated to be the very first Fullerian Professor in Chemistry within the Royal Organization.

Faraday was admitted to the American Academy of Arts and Sciences as an Honorary Foreign Member in 1832.

In in 1838, his name was appointed to join the Royal Swedish Academy of Sciences as an outside member. He was an associate member of the American Philosophical Society in the year 1840. In the year 1844 the American Philosophical Society was one of eight foreign members of French Academy of Sciences. He was appointed Associate Member from the Royal Institute of the Netherlands in the year 1849. which became known as the Royal Netherlands Academy of Arts and Sciences two years later. He was later referred to as an international member.

Faraday was afflicted with a nervous breakdown in 1839 but he was able to recover and resume his research into electromagnetism.

[2727 Faraday got a favour and grace home at Hampton Court, Middlesex, in the year 1848 as a result of the Prince Accompaniment's claims. The house was free of any costs or care. It is the master Mason's Home that was renamed Faraday Home and is presently being referred to as No. 37 Hampton Court Roadway. Faraday relocated to the area in 1858 in order to retire.

Faraday did not want to be part of the development of chemical weapons to be used during the Crimean War (1853--1856) after providing other projects of service to the British federal government. He identified moral issues that influenced his decision to not participate.

[29]

Faraday died on August twenty-fiveth of 1867 aged 75, in his residence at Hampton Court.

[30] He'd previously reduced the number of burials within Westminster Abbey on his death and was honored by a memorial plaque close to Isaac Newton's grave.

Faraday was interred in Highgate Cemetery West's Dissenters' (non-Anglican) area.

Chapter 3: Electrochemistry

Faraday began his career in chemical science working as an assistant chemist Humphry Davy. Faraday was particularly fascinated by the research studies on chlorine, and found two new combinations of chlorine and carbon. He also carried out the very first fundamental tests on gas diffusion that was first noticed through John Dalton. Thomas Graham and Joseph Loschmidt expanded the significance of this incident. Faraday had success melting various gases, investigating steel alloys, and creating several new types in optical glasses. A sample of one of these glasses was found to be large after Faraday computed the angularity of the plane of light's polarisation when the glass was inside an electromagnetic field. It was also the first substance that was discovered to be pulled back by magnet's poles.

Faraday created an early version of what would become the Bunsen burner which is still used as a source to heat for research centres all over the world.

[32]

[3333 Faraday was a chemist that discovered chemical compounds such as benzene (which he described as a bicarburet made with hydrogen) and also melted gases such as chlorine. The liquefaction of gas added to the idea of gases as the evaporates from liquids that have a very low boiling point and giving a more solid basis for the concept of molecular aggregates. Faraday first reported the synthesizing of carbon and chlorine compounds, C_2Cl_6 as well as C_2Cl_4 in 1820. He published his findings in 1821. Faraday also deduced the structure and chemical composition of chlorine clathrate hydratethat Humphry Davy had found in the year 1810. [3838 Faraday is also believed to be responsible for formulating the electrolysis guidelines and also promoting terms such as cathode and anode as well as electrode and ion. These terms were first suggested in 1810 by William Whewell.

Faraday was among the first scientist to define metal nanoparticles as they're currently referred to. In 1847, he observed that gold colloids possess distinct optical characteristics from the bulk metal equivalent. This was most likely the first discovery of quantum effects and could be considered to be the beginning of the field of nanoscience.

A chemical reaction occurs when the chain reaction is impacted by a potential distinction, like electrolysis, or when an electrical potential arises through a chain reaction like in a battery or fuel cell. Electrochemical responses, in contrast those of chain reactions do not transfer electrons (and consequently Ions) directly between particles, instead, they do so by using the previously mentioned electrochemical and ionically generating circuits. Electrochemical responses are distinct from chain reactions due to this distinction.

Electrical energy was first recognized during the 15th century. William Gilbert, an English researcher was a 17-year researcher in the century that followed to study magnetism and in a smaller way electricity. Gilbert was referred to as"the "Dad of Magnetism" for his research on magnets. He discovered how to improve and create magnets in a variety of ways.

Otto von Guericke, a German scientist, invented the first generator that was electrical in 1663, which generated fixed electrical power by applying tension to the device. A massive sulfur ball was placed inside the glass world, and then placed on a shaft that would later be the generator. The ball was turned by the crank and as a piece of paper put against the ball while it moved, an electrical trigger was generated. The planet could be out of the way and used as an electrical source to conduct experiments.

Charles Francois de Cisternay du Fay du Fay, a French scientist, discovered two types of electrical power that were fixed during the 18th century and found that charges of similar nature drive away each other, opposite charges draw them. Electrical power, as per Du Fay, is made from two substances: "vitreous" (from the Latin for "glass"), or positive"vitreous," also known as positive "resinous," or negative electrical power. This was the idea of two fluids of electrical power, but it would later be challenged later in the 20th century by Benjamin Franklin's one-fluid theories.

In his research into the laws of electrical repulsions, according to Joseph Priestley in England, Charles Augustin de Coulomb came across electrostatic attraction's law in 1785.

[5]

Luigi Galvani, an Italian physician and anatomist, discovered an association with electrical energy through chain reactions in his work "De Viribus Electricitatis in Motu Musculari Commentarius" (Latin for Commentary on the Impact upon the Influence of Electrical Energy on Muscular Movement) released in 1791. In it, he suggested an idea of a "nerveo-electrical substance" on natural living things.

[6]

Galvani wrote in his paper that animal tissue was an organic living power that he described in the essay as "animal electrical power," which stimulated nerves and muscles which were contacted by probes made of steel. Galvani believed that this new power, along with that of the "natural" form created by lightning, electrical eels and torpedo ray as well as that of the "artificial" form produced by friction, was a form electric energy (that is the fixed electrical power).

Galvani's medical colleagues mostly agreed with him, but Alessandro Volta belittled the concept of an "animal electrical fluid," declaring that the legs of the frog were responsive to differences in metal shape, mood and size.

Galvani resisted this by using two pieces of exactly the same substance to create muscles.

Electrolysis was employed in the work of William Nicholson and Johann Wilhelm Ritter in 1800 to convert water into oxygen and hydrogen. Ritter discovered the method of electroplating within a short time. He also noticed the spacing between electrodes affected the amount oxygen and metal that was transferred in the course of an electrolytic procedure. In 1801 Ritter had discovered thermoelectric currents, and was anticipating Thomas Johann Seebeck's invention of the thermoelectric.

Galvanic cells were developed through William Hyde Wollaston in the 1810s. Sir Humphry's research into electrolysis led to the conclusion that the creation of electricity in the basic electrolytic cells is due to chemical reaction, and that a chemical mix between two compounds occurred. In 1808 this research led to the separation of potassium and salt from their constituents, as well as the separation of earth metals that are alkaline from theirs.

Though he did not leave his subsequent studies on electromagnetism to other researchers, Hans Christian rsted's discovery of the magnetic effect of electrical currents in 1820 was immediately recognized as a landmark discovery. Andre-Marie Ampere without delay reproduced the experiment of rsted and mathematically outlined the results mathematically.

In the event of a thermal gradient in the joint, German-Estonian scientist Thomas Johann Seebeck revealed the electrical potential of the point sites of two different metals in 1821.

[12]

Georg Ohm, a German scientist, published "Die galvanische Kette, mathematisch bearbeitet" (The Galvanic Circuits Analyzed Mathematically) in 1827. He laid out his in-depth idea on electrical power.

Michael Faraday's research led to the solution to his two electrochemistry laws in 1832. John Daniell designed a main cell in 1836 which resolved the issue of polarization by stopping the development of hydrogen gas on the electrode that was positive. The zinc amalgamated could be re-aligned with mercury in order to generate an increased voltage, according to subsequent studies.

Decrease and Oxidation

"Redox" is the term used to describe "redox" describes the reduction-oxidation process. It refers to electrochemical processes where electrons are transferred between or away from the ion or particle which causes the state of oxidation for the particle. The reaction can occur when an external voltage is applied as well as when energy from chemical sources is released.

The change in the state of oxidation that takes place in molecules, ions or particles participating in the electrochemical reaction is known as an oxidation reduction. The theoretical charge an atom could have in the event that the bonds that bound atoms with various aspects were 100 percent ionic is known as the oxidation state. The state of oxidation of an atom or ion which loses electrons to another atom or ion increases the oxidation status that is the one who receives electron reductions that are negatively charged.

When atomic salt is integrated with the atomic chlorine, for example salt releases one electron, and then attains an oxidation state of 1. Chlorine takes in the electron, and reduces the state of oxidation to one. The amount of an electron's charge is linked with the signification of its oxidation state (positive/negative). The different charges of chlorine and salt ions draw each other, creating an inter-ionic connection.

Oxidation refers to the removal of electrons from an atom , or particle, whereas decrease is the increase of electrons. Mnemonic devices can aid in remembering the information. "OIL WELL" (Oxidation Is Loss, Decline Is Gain) and "LEO" the lion says "GER" are 2 of the most well-known (Lose Electrons: Oxidation, Gain Electrons Decrease). Oxidation and decrease typically occur in sets in which one species is burning while the other reduced. When electrons to be part of the same bond (covalent bonds) between atoms that have substantial electronegativity variations The electron is assigned to the atom having the highest electronegativity, when making the determination of the oxidation state.

The element that absorbs electrons is called the representative that oxidizes also known as an oxidant. the particle or atom which loses electrons is referred to as the representative that is lower or reducer. In response an oxidizing response, the representative of the oxidizing reaction is always reduced, whereas the representative that is reducing is in the process of being oxidized. The most common oxidizing agent is oxygen, however it's not all-inclusive. Whatever its name an oxidation response may not necessarily require oxygen. Fires that are caused by fluorine tend to be unquenchable due to the fact that fluorine is a stronger an oxidant (it has more electronegativity and thus can absorb electrons faster) as opposed to oxygen.

The process of oxidation that occurs in the particle or atom that oxygen has been added can be described in terms of the increase of oxygen during processes that involve oxygen (and the oxygen level is reduced). In natural particle like ethanol and butane suggest the oxidation of the particle where it disappears (and the hydrogen levels are reduced). This is due to hydrogen providing an electron for non-metals via covalent bonds. However, it also takes electrons with it when it's removed. Loss of oxygen , or gain in hydrogen however indicates a decline.

Cells Electrochemical

A cell that is electrochemical can be described as an instrument that generates an electrical current by releasing energy via the spontaneous redox reaction that can be triggered by electricity. The Galvanic Cell, also called the Voltaic Cell, was named for researches Luigi Galvani and Alessandro Volta who carried out various research studies into chain reactions and electrical phenomena in the late 1800s.

Two carrying electrodes are employed for electrochemical cell (the anode as well as the cathode). The anode is the electrode which undergoes oxidation. While the cathode is the one which undergoes reduction. Any substance that can effectively conduct electricity that is able to carry out the process, such as but not limited to, for instance, semiconductors, metals graphite, as well as polymers that conduct electricity, could be utilized to create electrodes. The electrolyte that contains free-moving ions, lies between the electrodes.

The galvanic cell is comprised of two distinctive metal electrodes that are covered in an electrolyte containing positively charged ions comprised of the electrode's converted form. A single electrode (the anode) will undergo oxidation, and the other reduced (the anode). The anode's metal will undergo oxidation, moving from a condition of zero to an optimum state of oxidation, and is an Ion. The metal ion at resolution takes in a few electrons from the cathode near the cathode, which reduces the state of oxidation to zero. This causes a significant transfer of the metal onto the cathode by electrolysis. The two electrodes must be electrically connected, allowing electrons to move from the anode's metallic layer to the ions that are on the cathode's surface by this connection. The electron flow can be described as an electrical charge that could be used for tasks such as turning a motor, or lighting a lamp.

Batteries

A variety of battery types have been marketed and are an essential use of electrochemistry in daily life.

[27] The first telephone and telegraph technology was powered with wet cells which also functioned as a source electroplating. The dry cell made of zinc-manganese dioxide was the first non-spillable portable battery that allowed flashlights and other gadgets that could be carried around to be operated. Mercury batteries, which made use of zinc and mercuric dioxide to provide more power and capacity than the dry cell originally used for early electronic gadgets is now out of widespread use because of the possibility of mercury contamination from disposal of batteries.

The lead--acid battery was the very first primary (rechargeable) battery that had external sources of replenishment. The electrochemical response that was the existing battery was changeable (to some degree) that allowed the energy of chemical and electrical to be switched off depending on the need. Sulfuric acid and water as well as lead plates, make up the majority of the lead-acid batteries. The most popular mix of acid is 30 percent. A problem is that, when the battery is without charge, the acid will form within the lead plates and render it ineffective. In normal use, lead acid batteries can last for around 3 years, however it's not unusual for an acid lead battery to be functional for 7 to 10 years. Lead acid batteries are often utilized in cars.

Electrolytes made of water are utilized in all the previous categories, limiting the optimal potential voltage for each cell. The low temperature efficiency is limited by frozen water. The lithium battery is superior to others due to the fact that it does not (and isn't able to) utilize water as the electrolyte. Rechargeable lithium-ion battery is a vital element in many mobile phones.

Because its reaction components could be pumped out by tanks outside and it is which is a variant that is speculative, has a significantly higher capacity for energy. It converts the chemical energy contained in hydrocarbon gases , or hydrogen directly into electrical energy, with more efficiency than any combustion process. these devices have power for many spacecraft, and are now being utilized to reduce grid power consumption for the public energy system.

How Can It Be Used?

The finishing [34] of items that have metals or oxides by electrodeposition, the adding (electroplating) or removal (electropolishing) or removal of thin layers metallic material from the surface of an object and the recognition of alcohol in intoxicated motorists via the redox response that ethanol produces are vital electrochemical processes in the natural world and in the market. The process of making metals like titanium and aluminum out of their mineral ores an electrochemical procedure, as is the production of energy from chemical reactions by photosynthesis. Certain blood glucose meters for diabetes utilize the redox potential of blood to measure the level of glucose present in the blood. A wide range of emerging electrochemical inventions, such as fuel cells, large size lithium-ion batteries in large format, electrochemical reactors and super-capacitors, are significantly industrial, and are in addition to earlier electrochemical breakthroughs (such such as lead acid deep-cycle batteries). Electrochemistry has also been found to be essential for the food industry such as determining the free acidity levels within olive oils, assessing the interactions between food and packaging, studying the structure of milk, characterization and determining the freezing point of ice cream blends and determining the freezing point of ice-cream blends.

Chapter 4: Electromagnetism

His work in the areas of magnetism and electrical energy is well-known. A voltaic stack made up of seven British halfpenny coins, seven disks of zinc sheet, and six sheets of paper soaked in seawater was his first well-known experiment. He degraded magnesia sulfate using his pile (first note to Abbott on July 12th, 1812).

Davy along with British scientist William Hyde Wollaston tried, but failed, to create an electric motor in the year 1821. This was shortly after Danish scientist and physicist Hans Christian rsted found the phenomenon of electro-magnetic.

3. After discussing the subject with the two guys, Faraday continued to build two devices to achieve "electro-magnetic rotation." The first, later called the homopolar motor generated a constant, circular motion through the creation of the force of a circular magnetic field in the

form of a wire that was extending into a mercury pool comprised of a magnet. If powered by a battery made of chemical then the wire would concentrate around the magnet. The foundation of current electro-magnetic invention was established through these experiments and breakthroughs. In his excitement, Faraday released his discoveries without mentioning his collaboration along with Wollaston as well as Davy. Faraday's coaching relationship with Davy was damaged due to the ensuing argument that could be a contributing factor in addition to his other commitments and prevented Faraday from conducting the field of electro-magnetic research for several years.

Faraday continued to work in his lab following his initial discovery in 1821. He began to investigate the magnetic properties of metals and developing the required capabilities. Faraday constructed a circuit in 1824 to determine whether electromagnetic fields could regulate the flow of electrons within a neighboring wire however, he could not find a connection. The experiment was a follow-up to three years of work similar to this using magnets and light, which produced similar results. [4646 Faraday continued to work for the next seven years perfecting his recipe to make optical high-quality (heavy) glass called borosilicate of lead, which would later be used in his research on the connection between to magnetism and light. Faraday released his speculations on electromagnetics and optics in his spare time and he also referred scientists whom he met during his travels across Europe along with Davy and also studying electromagnetism. Davy began his huge collection of experiments in which Electro-magnetic Induction was discovered two days after his demise in 1831. He was noting in his laboratory note pad on the 28th of Oct. 1831 that Davy did "doing different try outs the great magnet of the Royal Society."

When Faraday made two wire coils that were insulated within an iron ring, and carried a current through one of the coils it was discovered that a current of a short duration was produced in the other coil.

Shared induction is the name used to describe this phenomenon. In the Royal Organization still has the iron ring-coil machines on display. He observed that moving a magnetic rod through an unidirectional wire caused an electric current to flow through the wire later in the tests. If it was relocated on an unfixed magnet, the existing streams of electricity also flowed. The results of his experiments showed the idea that changing electromagnetic fields generates an electrical field. this connection was mathematically formulated as Faraday's Law from James Clerk Maxwell, which evolved into one the four Maxwell formulas and has evolved into the generalization of field theory. Faraday was to design his electrical anxious beaver that is the leading model of modern power generators and electric motors, which are based on the theories he discovered.

Faraday used "fixed," batteries, and "animal electrical energy" to cause electrostatic tension, electrolysis magnetism and other phenomena during 1832 as part an experiment aimed at understanding the fundamental characteristics of electricity. He concluded that, in contrast to the common perception at the time the differences between the various "kinds" of electrical power were not real. In fact, Faraday suggested that there's only one "electrical energy," and that different amounts and strengths (present in the form of voltage and present) result in different groupings of phenomena.

Faraday suggested near the end of his professional career that electro-magnetic forces extended into the space around the conductor.

[52[52 Faraday was not able to have his ideas accepted by the medical group, while his concept was rejected by colleagues in research. The idea of lines of flux that derived from magnets and charged bodies created a graph of electromagnetic and electrical fields. This model was crucial to the development of electromechanical gadgets that guided manufacturing and the market for the rest of the nineteenth century.

Diamagnetism

Faraday discovered in 1845 that a lot of materials showed some repulsion towards an electromagnetic field. This was referred to as diamagnetism.

[55]

Faraday also discovered that, by applying an electromagnetic field externally aligned with the direction the light is making to travel, the aircraft of the linearly polarized light could be altered. It is called the Faraday Impact is the term used to describe this phenomenon. "I have at last succeeded in brightening a magnetic curve or line of force and in magnetising a ray of light," He wrote on his notepad during the month of September 1845.

Faraday later utilized an spectroscope in search of an entirely different type of light-related modification which was the alteration of the spectral lines caused by an electromagnetic field in 1862. But the technology that he developed was insufficient to produce a conclusive study of spectral shift. Pieter Zeeman later studied the exact same phenomenon using an even better instrument and released his findings in 1897 and receiving the Nobel Reward in Physics in 1902 to recognize his achievement. Zeeman addressed Faraday's research in both his paper from 1897 as well as his Nobel acceptance speech.

"Faraday Cage "Faraday Cage" is a idea that has been in use for quite some period of.

Faraday's ice-pail experiment revealed that the charge was able to remain on the surface of the charged conductor. The charges from outside had no impact on anything contained within the conductor. This was proven in his research on static electrical power. This is due to the fact that outside charges disperse, causing the inside fields that emanate from them, causing them to oppose. The current term is a Faraday cage this protective effect is not utilized. Faraday constructed a 12-foot square wood frame that had four support structures made of glass, paper walls and wire mesh during the January of 1836. He went inside, and turned on the power. Faraday was able to prove that electrical power is the result of a force, not an impermanent fluid, after he came out of his cage.

Chapter 5: Civil Service and the Royal Organization

The Royal Organization of Great Britain was in a long-term connection with Faraday. In 1821, he was appointed assistant superintendent of the Home of the Royal Organization. In 1824, he became to be a member of the Royal Society. He was named as the Director of Royal Organization's Lab in 1825. Faraday was named one of the very initial Fullerian Professor in Chemistry within the Royal Organization of Great Britain six years later, in 1833, an appointment to which was appointed for all time, and without the requirement to deliver lectures. John "Mad Jack" Fuller who was the person who created Faraday's position in the Royal Organization, was his patron and coach.

Faraday performed various long, but usually lengthy, services for the economic sector along with his British federal government, in addition to his research studies with the Royal Organization in parts like electrical power, chemistry, and magnetism. The job involved examining blasts in coal mines, proving that he was a competent witnesses in the courtroom, as well as making the highest quality optical glasses for Chance Bro' lighthouses, which he completed with two engineers working for Chance Bro c. 1853. In the year 1846 the author co-authored an extensive and thorough research study along alongside Charles Lyell on a major blast at a colliery situated in Haswell, County Durham, which caused the deaths of 95 miners. Their findings were a thorough investigation into the forensics of the blasts that revealed that coal dust was a factor in the power of the blasts. Faraday demonstrated how ventilation could halt explosions, for the very first time during an explanation of how to stop the blasts.

The newspaper must have informed coal owners of the dangers of dust blasts caused by coal and, up to when the Senghenydd Colliery Catastrophe danger was not considered for more than sixty years.

Faraday worked for a long time as a highly respected researcher in a nation that has many maritime interests. He was involved in issues such as the building and and operation of lighthouses as well as the defense against rust of bottoms of ships. His workshop, which was located over the Chain and Buoy Store and near London's unique lighthouse, where he conducted the first attempt at lighting systems for lighthouses using electricity, remains in use.

Faraday was also involved in the field of ecology or engineering, in the way it is nowadays known. He was able to check for commercial contamination at Swansea and was advised of the Royal Mint about air contamination. Faraday was a witness to a letter addressed to The Times in the month of July 1855 regarding the state of the River Thames, which led to a widely circulated caricature appearing in Punch. (Likewise see The Great Stink).

Faraday was involved in the preparation and judging of exhibits in the city's Great Exhibit of 1851.

[64Then he was a member of the National Gallery Site Commission in the year 1857. He also was a consultant to on the National Gallery on the cleansing and maintenance of its collection of art. One of the other aspects of his duty was education. He presented a lecture on the subject in 1854 to the Royal Organization, and in 1862, he was appositive in front of the Public Schools Commission about education in the U.K.. Faraday also critiqued the public's obsession on table-turning, mesmerism, and seances, and chastised at the common public as well as the academic system in the same breath.

Faraday taught chemistry at The City Philosophical Society from 1816 to 1818 in order to increase the value of his talks prior the popular Christmas lecture he gave.

[71] Faraday conducted 19 Christmas-themed lectures for young people within the Royal Organization in London between 1827 between 1827 and 1860, a tradition that continues to be held today. The Christmas lectures of Faraday were designed to expose science to the general public with hopes of energizing them to earn a wage to the Royal Organization. The events were attended by a large portion of the London's upper crust. Faraday explained his views about learning in various letters to his colleague Benjamin Abbott: "A flame must be lit at the start

and preserved alive with constant luster to the conclusion," Faraday declared. His lectures were easy and energetic, including creating soap bubbles by filling them with various gases (to determine whether the bubbles were magnetized) before his audience members and marveling at the vivid colors of polarized light, however they were also incredibly philosophical. The audience was challenged to consider the physics behind his experiments during the lectures: "You're aware that ice drifts on water ... What triggers the ice to drift? Consider it and philosophize ". 1841 The Aspects of Chemistry 1843 The First Concepts of Electrical Power, 1848 The Chemical History of the Candle Light, 1851 The Chemical History of a Candle Light, 1851 Appealing Forces 1853 Voltaic Electrical energy 1854, The Chemistry of Combustion 1855 The Distinct residences of the typical metals 1857 Distinct characteristics of the typical metals 1858, Fixed Electricity 1859 The Metal Residence and the Relationships Between the various Forces in Matter.

An Unconventional Path

"But still try, for who knows what is possible..." – Michael Faraday, *The Life & Letters of Faraday, Volume II* (1870)

Unlike the scores of scholarly minds fortunate enough to be born and bred in the lap of luxury, few, if any, forecasted greatness or even a glimmer of eminence in Michael Faraday's future. Born on the 22nd of September, 1791, in the dusty farming village of Newington Butts, Surrey, in central London, the child, though adored and welcomed by his parents, further tightened the family's already strained budget. His father, Yorkshire-born James Faraday, was a sickly blacksmith who had been forced to part with his beloved smithery in Outhgill, Cumbria not long before the birth of his third child. James, a casualty of the plummeting economy triggered by the French Revolution, had no choice but to uproot his entire family, including his wife, Margaret Hastwell, along with his two children, Robert and Elizabeth. Margaret, the daughter of a farmer, had been employed as a meagerly paid servant girl before transitioning into a homemaker, and she was a loyal wife who supported her husband throughout all his endeavors.

They initially sought out the help of a fellow church member, another ironworker by the name of James Boyd, to help them settle in to their new home. Boyd provided James with a temporary job as an extra hand at his workshop, as well as a modest shack in Gilbert Street, Mayfair. Alas, the ailing James found it difficult to keep up with the steep cost of living in the area, so in the spring of 1791, he moved his family once again to Jacob's Well Mews, George Street, in what is now Westminster of the Greater London area. It was in the dingy second story of this ramshackle house, capped with a leaky tiled roof and built over a commercial stable, that young Faraday spent the bulk of his childhood.

James naively believed that the solution for his financial woes lay in the bustling and vibrant city of London. Back in Yorkshire, neighbors who could afford a trip to the grand city were regarded as local celebrities of sorts; home-bound friends and villagers often pressed them about the sights, most especially curious about whether or not the city's streets were truly "paved with gold," as the rumors said. As such, James pounced on the opportunity to cash in on the bottomless jobs that awaited him in London.

Much to his dismay, while the city exuded life and promise, no one was looking to hire a man who carried with him such baggage. Though only 30 years of age, James' frail body had taken a severe beating from his multiple illnesses, rendering him incapable of strenuous physical labor or a steady schedule, and therefore ineligible for almost all of the jobs available to a man of his limited education and skill set. When even workshops and smitheries began to turn him down, James could only turn to his irregular and below-minimum-wage job as a breadmaker at Welbeck Street to provide for his still-growing family.

As a result, young Michael wore ill-fitting hand-me-downs that belonged to his older brother, and seeing as how his parents could scarcely afford the roof over their head, he was made to play with the often broken toys that his older siblings had outgrown. That said, he was mature for his age and appeared to understand his family's situation, rarely giving his parents grief for what he knew they could not provide. The child was often spotted in the streets, particularly in Welbeck, where his father worked, typically armed with an incomplete set of marbles that he played with for hours on end. At times, he accompanied his younger sister, Margaret, who was 7 years his junior, to what is now Manchester Square, so that his exhausted mother could sleep for a wink or two.

Responsibilities aside, young Michael was similar to boys his age in plenty of ways. His thirst for excitement and urge to test the boundaries got in the way of his better judgment on more than one occasion. An account relayed by one of his sisters tells of the time he lost his footing when playing on the second story in his father's place of employment. As a result, the child plunged through the dry rot on the brittle wooden floorboards and may have lost his life had his father's station not been directly below him, ultimately softening his fall. The boy rolled off his father's back and pleaded for his father's forgiveness at once, but other than being shaken for a few minutes or so, he seemed oblivious to his close call with death, and he was back to his fun-loving self just an hour later. Some say it was his fighting spirit, inquiring nature, and ability to bounce back – traits he seemingly possessed even at a young age – that fostered his capacity for the mind-numbing trial and error process that comes with experimentation.

By the end of the 18th century, the family had sunk even deeper into poverty. Prices, including those of basic staples, continued to climb, but wages remained stagnant. At one point, farmers were peddling a quarter of a bushel of corn for £9 (roughly £709 - £900 in purchasing power today), instead of the usual 40 shillings. To put this into perspective, the annual incomes of lowly laborers in London averaged no more than a trifling £20. One in James' position was considered lucky if he could pocket 6 shillings for a week's worth of honest hard work. During the darkest of days, Michael and his siblings had no choice but to subsist on a single loaf of bread each for a week.

Despite the hardships at home, the one thing that traumatized young Michael the most was the inhumane treatment he often received simply because of his class status. This became a point of resentment that simmered within him until adulthood, and the speech impediment he had developed at a young age did little to improve his standing amongst his peers. His classmates at Sunday school, where he learned the basics of reading and writing, mocked him for the difficulty he had in pronouncing his Rs, and they cruelly referred to him as "Fawaday" both behind his back and to his face. Michael eventually made the acquaintance of a kind, elderly fellow from Cockney who sympathized with his plight and agreed to coach him a few times after school. One of the exercises his Cockney speech tutor often implemented during their private sessions required Faraday to keep his eyes riveted on the movements of his tongue when he spoke. This simple, but effective method is a technique still practiced by speech therapists around the world today.

Contrary to what one might expect, Michael failed to make much of an impression on his teachers. He was studious and kept to himself, but at times he was worryingly quiet, partly due to insecurities revolving around his speech impediment. And though he was a decent student who completed the majority of his assignments in a timely fashion, he mostly floated by under the radar, a preference that he grew accustomed to from there on out. When asked about what it was that kept him afloat during these rough times, he once remarked, "I was a very imaginative [child], and I could believe in the Arabian Nights as easily as in the Encyclopedia; but facts were important to me, and saved me. I could trust a fact, and always cross-examined an assertion..."

Michael's love for new information blossomed into something of an obsession. Even at the age of 9, he began to jot down a collection of facts that intrigued him in a small pocketbook. His journal, brimming with passages from books and snippets from newspaper and magazine articles on nature, arts, and sciences, was later compiled into a book entitled *"Michael Faraday's Philosophy Collection of Various Articles, Notes, Events, & Accidents in the Scientific World, 1800-1809."*

According to most biographers, his family's worsening finances compelled Michael to abruptly cease his studies at the age of 13. Other sources, however, describe a disturbing incident that they say is the real reason behind his permanent retirement from formal education. In the autumn of 1804, the teenage boy was summoned to the office of his schoolmaster. Not only was he subjected to merciless taunts and chastised for his speech impediment, the frightened boy was allegedly whipped and pummeled with blows until he lay in a bruised and quivering pile in the corner of the schoolmaster's office. Fortunately, Michael's brother Robert, who was in the classroom next door, investigated the ruckus, and once he saw the state of his brother, hastened home to inform his mother at once. Minutes later, Robert returned with an understandably outraged Mrs. Faraday, who, upon peeling her son off the ground, castigated the schoolmaster for his horrendous abuse of power and vowed never again to allow her children to set foot on the premises.

Whatever the case, 13-year-old Michael was withdrawn from his studies and made to pitch in on the family's mounting bills. Some have actually credited his removal from school as being a key to his future success; for example, Joel H. Hildebrand noted in a 1963 speech at NYU, "How fortunate for civilization, that Beethoven, Michelangelo, Galileo and Faraday were not required by law to attend schools where their total personalities would have been operated upon to make them learn acceptable ways of participating as members of 'the group.'"

Eventually, Michael picked up any odd job that made itself available to him, starting out as a chimney sweep and errand boy for wealthy families in the neighborhood. About a month or two later, the enterprising teenager spied a flyer advertising employment on a storefront situated on 2 Blandford Street. Wasting no time, he knocked on the door of the local book dealer and bookbinder by the name of George Riebau, and gently requested to apply for the position.

At this point, the teenager had conquered his speech impediment, and while still struggling with the perplexities of puberty, he had become far more confident than he was as a young lad. Riebau, who prided himself as an excellent judge of character, was moved by the young man's enthusiasm for such a plebeian position and agreed. Even so, Michael would have to prove his commitment, and he was thus placed "on trial" for a year as a store assistant and delivery boy for newspapers. 19th century Irish physicist John Tyndall poetically describes this period in Faraday's life: "[Faraday] slid along the London pavements, a bright-eyed errand boy, with a load of brown curls upon his head and a packet of newspapers under his arm..."

This strenuous period in his life was anything but glamorous, but it instilled in him the value of hard work and persistence. As an adult, he was known to have made sentimental remarks whenever passing by a young page or news courier: "I always feel a tenderness for those boys, because I once carried newspapers myself."

Once he had graduated from his one-year trial at Riebau's, 14-year-old Michael was made an apprentice bookbinder and stationer. To the teenager's delight, the generous Ribeau refused to accept even a pence for the 7 years' worth of shadowing and training, insisting that Faraday's proven loyalty was payment enough. As Riebau himself put it on the contract both parties signed on October 7, 1805, "In consideration of [Faraday's] faithful service, no premium is given."

James was overjoyed with both his son's work ethic and the favorable opportunity presented to him. Indeed, his elation is documented in a letter he wrote to his Clapham-based brother in the spring of 1809: "Michael...is very active at learning his business. He has been most part of 4 years of his time out of 7. He has a very good master and mistress, and likes his place well. He had a hard time for some while at first going; but as the old saying goes, he has rather got the head above water, as there are 2 boys under him."

All the while, Michael's hectic work schedule did nothing to dampen his quest for knowledge. Riebau's past apprentices never read past the titles of the books they rebound, but Faraday, on the other hand, devoured every last word of the books assigned to him. He even went on to use his bookbinding skills to bind his own assortment of scientific notes.

The books he fatefully came across further deepened his fascination with nature and science. He was particularly taken with the phenomena of electricity and magnetism, an attraction sparked by the first passage he chanced upon in the *Encyclopedia* Britannica: "Electricity." The 3rd edition of the *Encyclopedia*, which explored the accomplishments and experiments of various scientists throughout history, became a second bible of sorts for Faraday.

Thanks to the *Encyclopedia*, Michael became a devotee of Luigi Galvani, a trailblazer in the field of bioelectromagnetics and a bold experimenter with a specialty in the electric reanimation or "galvanization" of dissected animal corpses. This was the same Galvani who later inspired Mary Shelley to pen the Gothic literary masterpiece *Frankenstein*. Michael was also a great admirer of 18th century chemist, physicist, and inventor of the

electric battery, Alessandro Volta.

Galvani

Volta

Jane Marcet's *Conversations About Chemistry* was another heavily dog-eared book that Faraday kept around. The book's unconventional style – a dialogue-heavy educational novel centered on a teacher and a pair of pupils – was incredibly refreshing for a self-learner such as Faraday. He also borrowed from this manual many ideas for the amateur experiments he began to conduct at home, and thanks to the bolded warnings printed on the bottom of Marcet's illustrated experiments, he was aware of the hazards that came with certain substances, and thus always made certain to handle the substances properly.

Marcet

Faraday's fondness of Marcet's work is demonstrated in a letter he wrote to Swiss physicist Auguste Delyaria in 1858: "I felt then that I have found resistance of its chemical knowledge, and firmly grabbed for it. Hence the reason for my deep respect for Ms. Marcet. Firstly, it brought me personally great joy and made a real

65

blessing, and secondly, she was able to open to young, ignorant, inquisitive mind the phenomena and laws of the immense world of natural science…You can imagine my delight when I personally got acquainted with Ms. Marcet. How often did I mention the past and compared it with the present, so often I thought about my first teacher and always believed it is a duty to send their work as an expression of gratitude. And these feelings will never leave me…"

Rather than chase after girls or consort with others his age, Faraday spent what little free time he had with his nose buried in bulky books. At the age of 15, he began to invest whatever was left of his paychecks on chemical samples and other materials needed to construct homemade experimental paraphernalia. Many of the experiments he attempted to replicate on a small scale derived from multiple other science texts he happened upon, such as Isaac Watts' *Improvement of the Mind,* Robert Boyle's *Producibleness of Chemical Principles*, and *Lyons' Experiments on*

Electricity, among others. Faraday later elaborated upon his itch for experimentation, recalling, "I made such simple experiments in chemistry as could be defrayed in their expense by a few pence per week, and also constructed an electrical machine, first with a glass phial, and afterward with a real cylinder as well as other electrical apparatus [sic] of a corresponding kind."

Since Faraday was only compensated for weekend and overtime hours, he regularly made do with no more than a penny in his budget after all the bills had been paid. This being the case, improvisation became a skill the resourceful young man all but perfected. Substituting copper plates on pence coins with plates made out of zinc instead, for instance, the teenager succeeded in creating his version of the galvanic battery. Several weeks later, he constructed a primitive electrostatic generator using lumber, discarded bottles, and other scrap materials he scrounged up on the streets. Riebau, who was struck by Faraday's tireless resolve, further stiffened it with encouraging words, and Riebau even donated or allowed him to borrow items from his shop, such as a block of amber that Faraday subsequently used in his experiments with static electricity.

Another trait that distinguished Faraday from his peers was his faith. Shortly after their marriage, James was inducted into the Sandemanian (formerly Glasite) Church. This unorthodox Protestant sect, erected by Presbyterian minister John Glas, was born upon the deposition of its founder from the Church of Scotland in 1730. In short, the Sandemanians aimed to resurrect the ancient traditions of the Christian Church, with their governing body staffed with elders and pastors and a belief system that hinged on the "word and spirit of Christ." Followers condemned the concept of adhering to an overly political hierarchy, asserting that Christians should only be guided by "apostolic doctrine."

James was merely following in the footsteps of his father and the rest of the

Clapham-Faradays when he joined the church, but his connections with the unpopular sect worsened his reputation in the eyes of the general public. Its "unrighteous" members were easy targets, partly due to their pitiful size — there were no more than 100 Sandemanians in all of London. Still, the Faraday family convened with a handful of other Sandemanians in the "ghettos" and other less-frequented parts of the neighborhood to worship each and every week without fail.

Having grown up in such an environment, young Michael became a devout Christian himself. John Tyndall, a close confidante of Faraday's, once made this remark about the God-fearing scientist's bizarre, but seemingly effective routines: "I think that a good deal of Faraday's week-day strength and persistency [sic] might be referred to his Sunday Exercises. He drinks from a fount on Sunday, which refreshes his soul for a week."

Several characteristics that shaped Faraday's attitude towards his work stemmed from values he absorbed from the Bible. Not only did his faith explain many of Nature's most cryptic mysteries, the Book of Job – his favorite book in the Bible – serves as a reminder of the inevitable fallibility of mankind. As such, throughout his career, Faraday made certain to always present the unfiltered outcomes of his experiments and findings, and when one of his contemporaries shined a light on any errors he made, whether they were consequential or trivial, Faraday took them in stride and did his best to prevent his ego from getting the better of him.

By the age of 20, Michael, who was by now in the seventh year of his apprenticeship, was deeply grateful for having been able to broaden his arsenal of knowledge through the countless books he read during his bookbinding years. Still, he yearned for more, and with his paltry earnings preventing him from pursuing a full-fledged education, he reached for anything that came his way. Whenever a science professor or specialist held free public lectures, Faraday would be in the front and center of the crowd, conscientiously filling his journals with notes. That said, since these lectures were so few and far between, and they often lacked interactive segments that allowed students to voice their doubts and opinions, they failed to provide Faraday with the challenge he so desperately sought.

Mercifully, Faraday soon stumbled upon a band of young, like-minded science enthusiasts whose mutual passion for the subject surpassed their fiscal means. Dubbing themselves the "City Philosophical Society," the group, formed in 1808 by a young man named John Tatum, intended to supply a somewhat structured education that allowed them a more solid grasp of various sciences. Every Wednesday evening, a crowd of 20-somethings filed into Tatum's living room, where his classes – complete with a general curriculum – on different scientific subjects took place. Tatum also permitted his students to raid his personal library as they pleased.

Faraday was strolling down Fleet Street when a flyer tacked onto the window of a tailor's storefront caught his eye. For just one shilling, the flyer trumpeted, one could sit in on a lecture about natural philosophy delivered by locally renowned scientist and philosopher John Tatum. As negligible as this fee might seem to the average person, particularly when compared to costly tuition fees, it was to Faraday, a luxury he could not afford. At the same time, it was a luxury he simply could not do without. He reached out to those around him for a loan, until Robert, who was by now a blacksmith himself, agreed to gift his younger brother the shilling he needed.

Between February 1810 and September of 1811, Faraday dutifully attended 13 of Tatum's lectures. His ears remained perked as Tatum deconstructed and examined concepts regarding chemistry, galvanism, electricity, geology, meteorology, theoretical and experimental mechanics, and other topics that had been excluded from the textbooks at his disposal. Tatum also dazzled his students with relevant experiments he performed in real time, thereby providing many of them with their first contact with laboratory equipment. Tatum's live demo of a voltaic pile – a working electrical "battery" composed of copper and zinc discs, and a sheet of cloth soaked in brine (electrolytes) – was the first time Faraday had seen the contraption firsthand. He later organized his notes and compiled them into a four volume set, with the first volume written in honor of his employer, Ribeau.

The typically reserved Faraday would get out of his shell when he was in the company of the Philosophical Society, and he felt a sense of acceptance and camaraderie that he had been deprived of in his younger years. He formed an especially tight bond with a Quaker scientist named Benjamin Abbott, who earned his bread and butter as a clerk at a city house. In addition to their fiery and stimulating exchanges on philosophy and other scientific theories, the pair collaborated on experiments in Faraday's kitchen.

Tatum's lectures were not the only opportunities Faraday came across. Monseigneur John James Masquerier, a French exile who found refuge from the French Revolution in London, was lodging in Riebau's guest house as a tenant when he met the inspiring young man. Faraday was equally impressed by Masquerier, a respected artist whose claim to fame was a stunning, albeit evidently controversial portrait of young Napoleon Bonaparte. Awed by the young man's drive, Masquerier lent Faraday his books, provided him with basic geometry lessons, and taught him the crucial art of drawing in perspective. In return, Faraday cleaned up after Masquerier and maintained the "black in his boots." Faraday later recounted, "Masquerier lent me Taylor's *Perspective,* a 4-to volume, which I studied closely, copied all the drawings, and made some other very simple ones, as of cubes or pyramids, or columns in perspective, as exercises of the rules...I was always very fond of copying vignettes and small things in ink; but I fear they were mere copies of the lines, and that I had little or no sense of the general effect of and of the power of the lines in producing it..."

The one man who quite literally handed Faraday the ticket that galvanized his career was William Dance, another frequent patron of the bookbinding shop. There are two conflicting versions about the details surrounding their initial meeting.

Once Faraday had completed his apprenticeship in early October of 1812, he was scooped up by another bookbinding shop owned by Henri De La Roche. In early 1813, Mr. Dance wandered into De La Roche's shop and requested that a stack of chemistry papers be bound for him. Days after the expected delivery date, the peeved Mr. Dance stormed into the shop, but before he could demand an explanation, he found the bookbinder hunched over his notes, which Dance had inscribed during the lectures of Cornish scientist, inventor, and electrochemistry expert Humphry Davy. Dance was apparently so affected by the sight that he brandished and offered Faraday four tickets to Davy's lectures at the Royal Institution on the spot.

Davy

Another account adds a few more obstacles in the aspiring scientist's path.

Shortly after his apprenticeship, Faraday authored a heartfelt letter to Joseph Banks, president of the Royal Society, and humbly appealed to him for any position in the scientific field. In his own words, he wished to "be engaged in scientific occupation, even in the lowest kind." It was only after he received a rejection letter which included only two words in its curt response - "No answer" - that a disheartened Faraday accepted the position with De La Roche.

Faraday, however, never lost contact with the man who gave him his start in life, and likewise, Riebau never forgot his most diligent apprentice. And so, without the knowledge of Faraday, Riebau passed onto Dance's son some of the notes the apprentice assembled during his time with the Philosophical Society. Dance was astonished by the young man's pluck, but only after Faraday recreated one of Davy's experiments in front of him did the normally fastidious intellectual invite him along to the famously exclusive Institution.

Changes

"It is right that we should stand by and act on our principles; but not right to hold them in obstinate blindness, or retain them when proved to be erroneous." – attributed to Michael Faraday

Not only was Faraday thrilled to be granted what was to him a once-in-a-lifetime opportunity, he was looking forward to taking a breather from his day job. He had little to no complaints about the pitiful pay or the tedious labor that went into bookbinding, but his new employer, an uncouth, stubborn, and crotchety character, was testing, to say the least. While De La Roche was, at the end of the day, a man of good intentions who very much appreciated Faraday's maturity, trustworthiness, and unrivaled zeal, he could not for the life of him comprehend why the young man was so intent on pursuing a career in science. After all, there was much prestige, but little money to be made in the field of science in 19th century England, whereas bookbinding, comparatively speaking, was a far more lucrative trade. Unlike Riebau, De La Roche could not understand why Faraday, who was masterfully trained in a valuable skill, was wasting his time on a gamble, a gamble made even less likely due to how "uneducated" the impoverished young man was. De La Roche, who was not one to be shy with his opinions, told Faraday exactly what he thought of his taking time off work to attend a science lecture, but the young bookbinder paid the condescension no mind, grabbed his tattered coat, and went on his merry way.

Meanwhile, 35-year-old Humphry Davy, soon to become the President of the Royal Society, was as charismatic as they come. The local legend was intelligent and imaginative, and he commanded attention in almost every setting imaginable. The scientist's auburn locks, penetrative gaze, and casual confidence only made him a more captivating presence. As artfully articulate and charming as Davy was, however, it was his remarkable mind that consistently filled every last seat of the Institution's amphitheater.

Many knew and idolized Davy for his self-experimentation, most notably his use of euphoria-inducing nitrous oxide (laughing gas), but it was Davy's work with chemistry and batteries that drew Faraday to him like raw iron cobalt to a magnet. The "electrolytic apparatus" Davy created piqued Faraday's interest like no other. Simply put, the device was a battery hooked up to "metallic electrodes." These electrodes were then dunked into substances filled with whatever compound one wished to "decompose," or break down into singular elements. By melting certain minerals, amalgamating them with mercury, and piercing through them an electric current, he succeeded in disengaging pure barium from baryte (barium oxide). He also managed to isolate calcium from quicklime (calcium oxide), magnesia from magnesium oxide, and strontium from strontium oxide.

Davy's tendency to elegantly weave poetry into his lessons was a wonderful bonus, but it was his penchant for often explosive live experiments that made his lectures such a hit. Many of those who attended his lectures the previous year often bubbled over the one demonstration that truly amazed them. It sounded simple enough – all Davy did was flick a hunk of metallic potassium into a tub of water – but lo and behold, the potassium began to fizz and bounce about the surface of the water until, without warning, out coughed a billow of lavender-tinted smoke and flames.

Much to the glee of Faraday and his classmates, they were eventually treated to a more theatrical variant of this experiment. As described by Faraday's notes, one time in June 1813, Davy wheeled on stage a model volcano he had constructed out of clay, sand, and rocks. Shiny, silvery chunks of metal potassium peeked out from the veiny cracks at the base of the volcano. The majority of the crowd knew what to expect, but what ensued stunned them to silence all the same. Once Davy plopped the potassium into the crater and emptied a flask of water into the crater and cracks, a burst of crackling fire shot out of the volcano, followed by a cloud of purplish smoke and dozens of sparks jetting off in all directions like a tabletop fireworks display.

All those present were, needless to say, gripped by Davy's enchanting presentations, but none were quite as enamored as Faraday. Other students later recounted how the 22-year-old Faraday sat up in his seat in the upper gallery whenever Davy strutted up to the podium. It would have taken a disaster of cataclysmic proportions to snap the young man out of his trance as he hovered over his notebook, his eyes darting back and forth furiously as he listened attentively to Davy and rhythmically scratched walls of jottings at the same time. Nosy students who peered over his shoulder caught glimpses of the detailed sketches of laboratory apparatuses drawn into the margins of his notebook.

The following, in Faraday's own words, indicates the effect Davy had on him: "[Davy] spoke in the most energetic and luminous manner of the Advancement of the Arts and Sciences. Of the connection that had always existed between them and other parts of a Nation's economy. During the whole of these observations, his delivery was easy, his diction elegant, his tone good, and his sentiments sublime."

Faraday was so awestruck that he eventually began a relentless campaign to get his idol's attention. When he learned that Davy had temporarily damaged his eyesight – a result of one of his nitrogen trichloride experiments gone awry – Faraday sent the scientist numerous letters in the hopes of nabbing the position of laboratory assistant. To set himself apart from what he assumed was a high number of applicants Davy had to consider, Faraday attached to his letter an exhaustive, 360-page thesis based on the notes he had taken from Davy's classes.

When Davy came across this crisp stack of notes, beautifully bound in goatskin leather, he was genuinely amazed by the organizational skills and talent it required to produce such a manuscript, written in immaculate penmanship and enhanced by detailed, true-to-life illustrations. More importantly, he was impressed by Faraday's insight, so much so that he invited Faraday in for an interview despite knowing full well he had no space for him in his laboratory. As one might expect, Davy was blown away by the likable young man's seemingly promising talent, but since he was severely strapped for cash, he had no choice but to grudgingly impart to Faraday the bad news. "Science [is] a harsh mistress," Davy mused before the visibly deflated Faraday, gently urging him to seek his fortunes elsewhere. "And in a pecuniary point of view but poorly rewarding those who devote themselves to her service..."

Despite the demoralizing advice, Faraday's scientific ambitions persisted, and thankfully, patience would reward him. A few months after the failed interview, one of Davy's lab aides was fired due to the negative publicity that arose following a violent scuffle the aide in question allegedly provoked. Days after that, a letter carrier strode up to Faraday's front door and handed him an envelope marked with the distinctive sleepy scrawl of Sir Humphry Davy. The letter read, "I am far from displeased with the proof you have given me of your confidence, which displays great zeal, power of memory, and attention...I will see you at any time you wish."

That same afternoon, Faraday resigned, hanging up his apron and stowing away his type holder, stamps, engravers, and other bookbinding tools for the last time. He felt the natural sting of wistfulness for ending this critical chapter in his life, and though he was taking a rather drastic pay cut – his starting salary was no more than 25 shillings a week ($101 USD in purchasing power today) – he was finally being granted the chance to answer his calling. His family was pleased to see the fire in his eyes every morning as he headed off to work on Albemarle Street, but they were also understandably concerned. Given that both of Faraday's parents – and most of his ancestors, too, for that matter – were modest blue-collar workers who had no choice but to forgo classical educations themselves, they fretted over his future and income, more so now that James was out of the picture. To give a better sense of the currency's value at the time, a tailor-made suit could set one back £8.

Undeterred, and with fame and fortune being the last items on his agenda, Faraday awoke every dawn with a bounce in his step. With this winning attitude, it wasn't long before he proved himself to be a stellar lab assistant. He was eager to learn, followed instructions to a letter, and provided priceless, out-of-the-box observations that either changed the direction or expanded the breadth of his mentor's thinking. He even followed along without protest when Davy suggested they give his ill-fated chlorine experiments – which had resulted in unruly explosions on four different occasions — another whirl. Like his mentor, Faraday was fearless when it came to experimentation, so much so that it often bordered on reckless. On top of the minor, but multiple injuries he sustained from various experiments gone wrong, a flying shard of glass once threatened to rob him of his eyesight.

At the same time, the relationship between them wasn't always rosy. In the autumn of 1813, Davy was commissioned to headline a "grand tour of the continent." He was issued a special passport from Napoleon himself, one that extended the same traveling privileges to Davy's wife, Jane, and two other people. Davy originally planned for a maid and a valet to accompany them, but when the latter suddenly dropped out, he asked Faraday if he would kindly undertake the duties until he could find someone else along the way to fill the post. Faraday practically worshiped the ground Davy walked on, but he balked upon hearing this frankly degrading request. It took much persuasion on Davy's part, as well as the tempting promise of broadening his horizons, for Faraday to finally agree to the task.

As a result, in mid-October of that year, Davy, Jane, her maid, and Faraday boarded the carriage-for-hire and set off on their journey, alighting in France, Italy, Switzerland, and parts of southern Germany. It was to Faraday the adventure of a lifetime. They experienced the sights and visited a number of prominent museums and educational institutions. In Italy, for example, the group marveled at the beauty of Vesuvius and inspected Galileo's first-ever telescope up close. Faraday was also introduced to a colorful bevy of illustrious scientists, such as Paris-based physicist André-Marie Ampère, and his idol Alessandro Volta in Italy, as well as natural sciences professor Auguste-Arthur De La Rive and Charles Gaspard in Geneva. Their mobile scientific exploits only came to an end in April 1815, following Napoleon's return from exile.

Ampère

The eye-opening scientific tour around Europe would have been a flawless experience had it not been for the insufferable Jane Davy and her pompous, holier-than-thou attitude. Jane made it no secret that she thought herself superior to anyone with a less-than-glamorous upbringing. Refusing to see Faraday as her husband's equal, she prohibited the young man from dining with them, and she constantly ordered him to sit with her maid and the driver in the front of the coach. To make matters worse, she is said to have thrown fits whenever Davy and Faraday attempted to discuss scientific matters at length, ensuring the attention would be shifted back to her.

At one point, Jane's mood swings became so intolerable that a fed-up Faraday came close to voluntarily vacating his post until an apologetic Davy reeled him back in. This may or may not have had anything to do with the raise he received – 30 shillings a week – upon his return to England. The 16% pay increase not only allowed him to better foot the bills, he was able to save up for his sister's boarding school fees.

While Davy and Faraday certainly shared an authentic bond over the years, the latter would soon find out why heroes are meant to be met only in fantasies, and never in the flesh. In late August 1821, Faraday was tasked with designing an experiment modeled after one devised by Hans Christian Oersted, a Danish scientist who had just discovered that electrical currents were capable of "deflecting" the needles of magnetic compasses. Rather than conduct a stale duplicate of Oersted's experiment, Faraday created an experiment and a theory on the 4th of September that would put even its muse

to shame.

Oersted

At first glance, Faraday's experimental contraption, which aimed to further clarify the relationship between magnets and electricity, was aesthetically underwhelming. To the right of a wooden board was a tin bowl of mercury with a tiny magnetic cylinder in its center; the top of the magnet was referred to as the "north pole," and the bottom the "south pole." Suspended above the bowl was a lengthy thread of wire, hanging from a hook; the hook was attached to a rod wrapped in copper wiring, which in turn, was connected to a battery that provided a positive current. One end of a second strand of copper wire was also hooked up to the battery, while the other end was submerged in the mercury.

When Faraday activated the battery, the wire above the magnet began to spin around it, continuing to do so until he manually deactivated the battery. He discovered that when the north pole of the magnet was facing upright, the rotation of the wire was anti-clockwise; the wire only spun clockwise when he flipped over the magnet. This enlightening discovery marked his first contribution to the field of electromagnetism, for never before had anyone succeeded in creating "continuous motion" from chemical energy. As such, Faraday had unwittingly developed the fundamental mechanics of the electric motor. Interestingly enough, Faraday, though electrified by his discovery of "electromagnetic rotations," appeared to have been oblivious to the potential applications of his discovery. A note from his journal about this experiment reads, "Very satisfactory, but make more sensible apparatus..."

Davy, on the other hand, recognized the profundity and promise of his protégé's findings, and he had already been stewing over the prospect of being eclipsed by his protégé's burgeoning shadow. Declarations made by people like Sir Henry Paul Harvey, who once proclaimed Michael Faraday to be Sir Davy's "greatest discovery," ruffled Davy's feathers, but it was all the attention Faraday attracted from this discovery that shoved him over the edge. Much to Faraday's horror, his mentor accused him of misappropriating the idea from the interim president of the Royal Society, William Hyde Wollaston, and swiftly launched a scathing smear campaign against him. Many believe Davy, in cahoots with another envious member, Henry Warburton, attempted to irreparably tarnish Faraday's reputation in a bid to dissuade the Society from considering Faraday as a member. Thankfully, Wollaston himself believed Faraday to be innocent, writing to him in a letter, "Sir — you seem to labor under some misapprehension of the strength of my feelings upon the subject to which you allude. As to the opinions which others may have of your conduct...if you fully acquit yourself of making any incorrect use of the suggestions of others, it seems to me that you have no occasion to concern yourself much about the matter."

Wollaston

Warburton, whose whispers about Faraday often drowned out the rest, was eventually made to withdraw his accusations and publicly undo the damage he had inflicted upon Faraday's previously pristine reputation, but Davy never did retract his statements. Nevertheless, Faraday was ultimately granted a coveted spot in the Society later that year. He was then promoted to lab director of the Royal Institution in 1825, and upon Davy's retirement two years later, succeeded him as the Institution's Professor of Chemistry.

Although Faraday's name was more or less cleared, Davy's treachery still took its toll on him, so much so that he intentionally diverted his focus to other areas of experimentation, keeping anything electrical at arm's length until Davy's death in 1829.

1821 proved to be a pivotal year for Faraday in more ways than one. Apart from his momentous discovery and his promotion to acting superintendent of the House and Laboratory on the 21st of May that year, he made the acquaintance of a lovely young woman named Sarah Barnard for the first time. Little is known about the early life of the arresting 21-year-old maiden who tantalized Faraday with nothing but her wit and sense of humor, and where exactly they met is still a matter of dispute. Some say he met Sarah, the daughter of an elder of the Sandemanian Church, at service, while others claim they met at Faraday's laboratory in Paul's Alley after Sarah made the wrong turn one summer afternoon.

Sarah's brother, Edward, is a character often mentioned in their love story. Edward and Faraday are said to have belonged to the "self-improvement essay writing" society. The pair were always friendly with one another, but it was Faraday's eloquent papers deriding the concept of love that prompted an amused Edward to approach him. Edward was so tickled by Faraday's unique perspective that he told his younger sister about him, inadvertently taking on the role of matchmaker.

However it was they met, Faraday, who once scorned the very idea of wasting time with something as pedestrian as love, found in her a lifelong companion. On the 5th of July, 1820, he professed his feelings for her in a letter and requested her hand in marriage. To his disappointment, Sarah, seemingly agreeing with her father that "love [makes] philosophers into fools," was spooked by the bold request and skipped out to Ramsgate, leaving the baffled Faraday hanging. Undeterred, Faraday traveled to Ramsgate in late July, remaining there and proving his commitment to her before she finally agreed to marry him.

Bearing this in mind, it's important to remember that Faraday, calculating as he was, mulled over the notion of marriage for years before he popped the question. His closest friends remember how he went off on a spiel whenever the subject of marriage came into conversation, weighing the pros and cons of a lifetime union to both bachelors and married men alike. The scale often tipped in favor of the latter, for the dictatorial Lady Davy often sprang into mind. Another poem, found scrawled into his personal journal, shined a light on his opinions regarding the matter:

"What is the pest and plague of human life?

And what is the curse that often brings a wife?

'Tis love...

The noble heart will ne'er resign reason,

The light of mental day,

Or idly let its force decline before the passions' boisterous sway..."

It was Sarah Barnard, "the pillow to [Faraday's mind]," who managed to undo this pessimistic side of him. The couple wedded on the 12th of June, 1821 at the Church of St. Faith, and they remained devoted to one another for the rest of their days. A month after their wedding, Faraday chose to air his sins and profess his faith before the Sandemanian Church. As to why he chose to wait until the age of 30 to do so, Faraday's enigmatic reply was always the same: "That is between me and my God." While they had no children of their own, they were loving parental figures who doted on more than 80 nephews and nieces, even taking in two of their nieces for a couple of years each. Not only was Sarah a competent individual who maintained the household and justly directed the "servants" of the Institution, she was well-respected by Faraday's contemporaries. She was often seen in the company of esteemed scientists, such as Justus Liebig and Christian Schoenbein, gesturing wildly and speaking with great confidence as they latched on to her every word.

"Nature is our kindest friend and best critic in experimental science if we only allow her intimations to fall unbiased on our minds." – attributed to Michael Faraday

Following his unfortunate fallout with Davy in 1821, Faraday steered his attention towards the less controversial field of gaseous experimentation. The most impactful of these trials transpired a couple years later, when the industrious experimenter tackled a problematic series of chlorine-related investigations.

First, chlorine hydrate was poured into a sealed glass tube reminiscent of a thick bendy straw. The shorter end of the bent tube was placed atop an open flame, thereby "cooking" the liquid chlorine, while the longer end was immersed in a tub filled to the brim with salted ice cubes. Faraday watched with apprehension as pale yellow vapor materialized and began to dance within the tube. Just moments later, the vapor was condensed into an "oily" golden-yellow substance in the part of the tube inside of the ice bath.

Faraday had created liquid chlorine, but not without putting himself in danger on dozens of occasions. The worst of this set of mishaps occurred in late March of that year, when Faraday accidentally shattered the tube, causing an explosion "so powerful...it [drove] the pieces of glass like pistol-shot through a window." 13 jagged shards threatened to strike Faraday square in the eye, but the responsive researcher, fortunately, ducked away in the nick of time.

John Aryton Paris, one of the scientists working in the vicinity at the time of Faraday's breakthrough, supposedly scolded him for his less-than-polished workspace and the wretched "impurities" the experimenter seemed to be concocting. Faraday must have felt some semblance of smugness when he scribbled the following note to Paris the next day: "Dear Sir – the oil you noticed yesterday turns out to be liquid chlorine. Yours faithfully, M. Faraday."

Paris and Faraday's previous naysayers congratulated the resilient researcher on his unprecedented triumph in the laboratory. Davy, apparently, remained the only sour grape among the bunch. As maintained by Faraday's biographers, Davy went on to "steal" Faraday's unpatented process to liquefy a number of gases, such as the muriatic acid gas, which only intensified the friction between both parties, now in an unofficial cold war with one another.

Pettiness aside, Faraday's production of liquid chlorine was vastly influential, particularly in the sphere of thermodynamics. In order to convert gaseous chlorine into its liquid form, energy – in this case, heat – had been transferred. Furthermore, he had unearthed a method to cool surroundings through the absorption of heat, therefore creating the principle of refrigeration. More significantly, he had proven that "permanent gases were merely vapors of liquids...with low boiling points." His technique of liquefying gases, which he later implemented on nitrous oxide, sulphuretted hydrogen, and ammonia, was one adopted by several leading scientists of his era, as well as of future generations.

Faraday's innovative genes are said to have been so potent that he often minted inventions even without essentially meaning to. One of the most unique was the party balloon. The idea of a balloon itself was nothing new, as paintings from as early as the 1300s show people at social functions wielding a floppy, inflated sack of air fashioned out of pig bladders as a decorative accessory, tossing them about as "balls." Faraday wasn't even the first to use balloons in a laboratory setting; Galileo, for example, used pig bladder balloons as an instrument to measure the "weight of air." But what Faraday did do was create the world's first rubber balloon in 1824. Hoping to better analyze the chemical element of hydrogen, he laid 2 cutouts of "tacky" rubber on top of each other. Before thumbing down the edges of the cutouts to "seal" them, he first coated the inside of his "balloon" with a layer of flour so that the pieces did not stick together completely, creating a sort of vacuum in the center. He then pumped his balloon full of hydrogen gas and noted that while the inflated sack had "considerable ascending power," the elusive gas often found a way out.

Less than a year later, rubber entrepreneur Thomas Hancock began to hawk rubber toy balloons. His product, sold as individual do it yourself kits, came with a "condensing syringe," as well as a glass vial containing rubber solution. Latex balloons were later introduced to the market in 1847 by London native J.G. Ingram, and balloons became mass-produced merchandise in the 1930s.

Faraday's innovative abilities were also spurred by his unfaltering spirit and admirable ability to find inspiration anywhere around him. In the spring of

1825, Faraday suddenly found himself inexplicably drawn to the gas canisters being delivered to the Institution by his brother, Robert, at that point an employee of the London Gas Company. More precisely, he was anxious to learn more about the transparent, almost fragrant liquid that lingered in the bottom of the canisters. Thus, in May of that year, Faraday took it upon himself to isolate the hydrocarbon now known as "benzene" from the "pyrolsis of whale/fish oil." To do so, he heated up the oil with oxygen inside of an eudiometer, an upturned, test-tube-like apparatus designed to measure the change in volume of gases after these mixtures have undergone chemical alterations. As the new compound he extracted contained a uniform amount of hydrogen and carbon, he christened it "carbureted hydrogen." Later, the chemist Auguste Laurent proposed the compound be named "pheno" after the Greek term *phainein* ("to shine"), owing to its luminescent quality.

Faraday, who founded the system of oxidization numbers, was reportedly so proficient at calculating melting and boiling points that the data he recorded regarding benzene was strikingly accurate, even by the standards of today. The multifaceted hydrocarbon is now a vital component in a varied array of modern products, from the manufacturing of rubber and dyes to the construction of phenol, therapeutic chemicals, and even explosives.

It was also during Faraday's work with benzene that he devised what is now described by experts as an early version of the Bunsen burner. Three years later, Faraday, purportedly motivated by Davy's miner's lamp, engineered a gas burner to be used in the laboratory. His prototype, a plain metal tube topped by a removable "chimney-pot," was quite straightforward – to crank up the heat, one had to move up the chimney-pot, and vice versa. Other Faraday-branded inventions still found in many laboratories today include the Faraday cup, a doorknob-shaped device used to measure particle currents, as well as Faraday flashlights.

In 1827, Faraday debuted his first-ever published book, *On Chemical Manipulation.* This piece received much approbation from his peers, many of whom applauded him for his fixation with giving credit where it was due. Just months before the publication of his book, Faraday put in circulation a short work about fluid sulfurs with the following footnote attached: "I have just learned that Signor Bellani had observed the same fact in 1813...Bellani complains of the manner in which facts and theories which have been published by him are afterwards given by others as new discoveries; and though I find myself classed with Gay-Lussac, Sir H. Davy, Daniell, and Bostock, in having thus erred, I shall not rest satisfied without making restitution, for M. Bellani in this instance certainly deserves it at my hand." Admissions like this are most likely why the majority of scholars found themselves siding with Faraday, as opposed to accusers such as Sir Davy.

In August 1831, about two years after Davy's death, Faraday initiated a dogged 10-day experimentation session that would solidify his name in the realm of electromagnetic research. For over six years, he had pondered whether or not it was possible for an electric current, when traveling through a conductor, to induce a charge in a nearby conductor. With his rancorous mentor now at rest, he was finally comfortable enough to put this to the test.

Months earlier, back in the springtime, Faraday had teamed up with Charles Wheatstone, an English physicist and versatile inventor best known for the electric telegraph and being the first to coin the term "microphone." Together, the duo began an intensive study on sound. One experiment involved sprinkling a chalky powder onto iron plates, which were then "awakened" by the vibrations from a violin bow. He found the hypnotic patterns that formed in the quivering powder, known as "Chladni figures," a curiously beautiful effect, but he was even more impressed by the fact that these figures could be aroused by "bowing" a neighboring plate. This "acoustic induction" was, to Faraday, conclusive proof that the same effect could be

replicated with electrical wires.

Wheatstone

On the 29th of August, Faraday unveiled a mechanism that aimed to do just that. The mechanism consisted of a soft iron ring, measuring about 6 inches across, bandaged with five coils of copper wire, with about 2 millimeters of space between each coil. Since insulated wire was extremely difficult to come by, the ever-resourceful Faraday "insulated" his wires with strips of cotton and calico that he had snipped off from Sarah's old petticoats. The opposite ends of the iron ring were linked to a pair of unconnected copper wires (four in total). One wire on each side was then linked to a voltaic battery, and the remaining wires were hooked up to a galvanometer, a device that measures electric currents. When the battery was switched on, the galvanometer's needle jolted to life, instantly detecting the current on one side of the ring. To his astonishment, the needle dropped back to 0, apparently detecting no current at all, notwithstanding the live status of the battery. Faraday discovered, however, that if he were to continually switch the battery on and off, he could recreate the effect as many times as he desired.

Faraday recorded the exciting effect in his journal: "Upon using the power of 100 pair of plates with this ring, thei mpulse at the galvanometer, when contact was completed or broken, was so great as to make the needle spin round rapidly 4 or 5 times before the air and terrestrial magnetism could reduce its motion to mere oscillations."

Through this phenomenon, now known as "electromagnetic induction," Faraday invented the world's first transformer, a revolutionary piece of technology that would one day spawn the likes of phone chargers and electricity substations. That said, while the inventor himself recognized the magnitude of his discovery, he remained uncertain about its diverse applications, and he would not live to fully appreciate them in their full glory. "I am busy no again on electromagnetism and think I have got hold of a good thing, but can't say," reads a passage from his journal. "It may be a weed instead of a fish that, after all my labor, I may at last pull up."

In 1832, his research on the subject was enhanced by another contraption of his creation. Upon realizing that he needed a new method of "producing a changing magnetic field," he decided to modernize Francois Arago's famous disc experiment. First, he placed two bricks of horseshoe magnets on metal platforms, and he then wedged between the magnets a copper disc that was hooked up to a galvanometer. When the disc spun, the galvanometer, as predicted, detected a current moving in a "radial direction." By reversing the rotation of the disc, the needle of the galvanometer jumped to the opposite side of the meter accordingly. "Here therefore was demonstrated the production of a permanent current of electricity ordinary magnets," read an excerpt from Faraday's journal, describing his work on the Faraday disc. "If a terminated wife is moved so as to cut a magnetic curve, a power is called into action which tends to urge an electric current through it." He had proven how a synchronized partnership between a magnetic field and "continuous mechanical motion" could generate an unbroken electric current, indirectly presenting to the world the first electric generator. But as groundbreaking as his Faraday's disc was, the mechanism was deemed far too unstable and volatile for it

to be converted into mass use.

A Faraday disc

Undeterred, Faraday continued with his research, diving deeper into the field of chemistry and its connections with electricity later that same year. On one side of the table was a massive glass test tube filled with platinum electrodes "dipped in molten tin chloride," boiling atop a controlled flame. These electrodes were linked to an analog voltmeter (a scale-like device that used oxygen and hydrogen gas to measure the difference between two electrical points in a circuit), as well as a voltaic battery. By weighing the coils, he was able to gather that the amount of tin deposited was equivalent to the quantity of electricity generated. This discovery, which would one day lead to anodes and cathodes, battery components now used in phones, flashlights, digital cameras, and other electronics, was rounded out by the two laws Faraday established shortly thereafter. The laws of electrolysis, as defined by Doug West in *Owlcation*, are as follows: "One – The amount of a substance deposited on each electrode of an electrolytic cell (in the form of ions) due to flow of current is directly proportional to the quantity of electricity (measured in coulombs) passed through it... Two – The mass of the

115

substances deposited when the same quantity of electricity is passed through several electrolytes are in the ratio of their chemical equivalent."

1832 proved to be yet another eventful year. Apart from Oxford University awarding him an honorary doctorate for his multiple advances in the world of science, Faraday was appointed deacon in his local Sandemanian church.

Faraday once confided in a friend that he functioned best when working in solitude. "I have never had any student or pupil under me to aid me with assistance, but have always prepared and made my experiments with my own hands...I do not think I could work in company, or think aloud, or explain my thoughts at the [same] time. Sometimes I and my assistant have been in the laboratory hours and days together, he preparing some lecture apparatus or cleaning up, and scarcely a word has passed between us..."

Be that as it may, Faraday felt indebted to those who had provided him with the education and training he needed to launch his scientific career, and he made it his life's mission to give back to the community. He was 24 when he delivered his first lecture on the different properties of matter to the members of the City Philosophical Society in 1816.

He was eventually appointed a lecturer at the Institution, and he later lent his services to Woolwich part-time. His résumé only continued to blossom, as he served as the Scientific Adviser to the Admiralty, and the educational part of his career culminated when he was appointed to the tenured position of Fullerian Professor of Chemistry in 1833.

Remembering his roots, Faraday also designed special syllabi for public lectures. He started the Royal Institution's Christmas Lectures in 1825, which, as its name suggests, covered various scientific subjects on the yearly holiday, as well as the Friday Evening Discourse, which allows different scientists and scholars a chance to disclose to the community the latest in technological and scientific innovations. These lectures, which are still held today, were shifted to the small screen in 1966, and they are now available online. Faraday always regarded science as a right, rather than a privilege, and one that should always be accessible to the masses regardless of history or social standing. "If the term education may be understood in so large a sense as to include all that belongs to the improvement of the mind," he argued emphatically during one lecture. "–either by the acquisition of the knowledge of others, or by increase of it through its own exertions, we learn by them what is the kind of education science offers to man. It teaches us to be neglectful of nothing – not to despise the small beginnings, for they precede of necessity all great things in the knowledge of science, either pure or applied."

An illustration depicting Faraday delivering a Christmas Lecture in 1856

Faraday, who once stuttered in the company of those outside of his immediate family, was determined to be as engaging and illuminating as the authoritative professors he studied when he was an ambitious student. Judging by the testimonials of those blessed enough to be granted a seat in Faraday's classes, he more than succeeded, with a passion so spirited that it overcame his stage fright and remaining inhibitions. Friedrich von Raumer, a liberal German politician who attended several of Faraday's lectures, once said that he "speaks with ease and freedom, but not with a gossipy, unequal tone, alternately inaudible and bawling, as some very learned professors do; he delivers himself with clearness, precision, and ability. Moreover, he speaks his language in a manner which confirmed in me a secret suspicion that I had, that a number of Englishmen speak it very badly." Another student named Jane Pollack shared this about Faraday, whom she presumed to be a natural-born speaker: "It was an irresistible eloquence which compelled attention and invited upon sympathy. There was a gleaming in [Faraday's] eyes which no painter could copy, and which no poet could describe. Their radiance seemed to send a strange light into the very heart of his congregation, and when he spoke, it was felt that the stir of his voice and the fervor of his words could only belong to the owner of those kindling eyes...His enthusiasm seemed to carry him to the point of ecstasy when he expatiated on the beauties of Nature, and when he lifted the veil from her deep mysteries...His audience took fire with him, and every face was flushed."

Further Creations and Legacies

An 1850s depiction of Faraday in his lab

An 1870 engraving of Faraday's lab at

the Royal Institution

An illustration of Faraday's study at the

Royal Institution

An illustration of Faraday's flat at the Royal Institution

"No matter what you look at, if you look at it closely enough, you are involved in the entire universe." – attributed to Michael Faraday

For the next few years, Faraday resumed his research on electrochemistry. In 1834, he attempted to tie his research together with new terms for the concepts he had contrived. To do this, he consulted respected polymath William Whewell of Cambridge, who was most renowned for having substituted the word "scientist" in the stead of "natural philosopher" for the first time in 1834. Whewell's response to Faraday's list of suggestions, dated May 6, 1834, reads as follows: "I still think 'anode' and 'cathode' [are] the best terms beyond comparison for the 2 electrodes." Satisfied, Faraday integrated these terms into his journal on the 13th of May, and Faraday's confirmation letter was mailed to Whewell a few days later. "I have taken your advice, and the names used are 'anode,' 'cathode,' 'anions,' 'cations,' and 'ions'; the last I shall have but little occasion for. I had some hot objections made to them here and found myself very much in the condition of the man with his son and ass who tried to please everybody; but when I held up the shield of your authority, it was wonderful to observe how the tone of objection melted away."

In 1835, Faraday began to grapple with the concept of static electricity. At this stage, his peers at the Institution were no strangers to the professor's antics, so they barely batted an eye when the whistling chap lugged a large metal ice bucket into the laboratory, a spool of silk thread, a wooden stool, and a metal ball tucked under his arms. First, he placed the "uncharged" bucket, which was to serve as a conductor, onto the wooden stool, so as to insulate the pail from the ground. He then lowered a "positively-charged" metal ball into the small opening on the lid of the bucket with a length of "non-conductive" silk thread, hoping to observe how charges reacted when another charged entity was introduced to the inside of the conductive bucket. With the help of a gold-leaf electroscope, which he had connected to the bucket, Faraday concluded that the negative charges were confined to within the bucket, leaving all the positive charges on the exterior of the enclosure. Moreover, when he lowered the ball even further, allowing it to make contact with the bottom of the bucket, the positively-charged ball neutralized the charges outside of the bucket, rendering both charges equal. The effect Faraday successfully demonstrated is now known as "electrostatic induction."

Faraday thus developed a theory now embraced as the "Faraday Cage Principle." Charges, he speculated, only survive on the outside of "live" conductors. Due to exterior charges automatically rearranging themselves, they would have no effect on anything confined within the conductor.

Naturally, another experiment was required to prove this theory, so in 1836, Faraday constructed a tiny room and layered on all six faces of the cubicle sturdy, scintillating sheets of metal foil. He then hooked up the room to an electroscope and zapped the facade of the cubicle with bolts of electricity discharged from an electrostatic generator. Much to his excitement, the electroscope indicated that not a single bolt had managed to pierce through his Faraday Cage.

The discovery of the principle and cage continue to bear fruit to this day, with seemingly endless applications that extend to all walks of life. The shells of cars, planes, and other closed vehicles, for instance, conduct themselves as Faraday cages of sorts, for the aluminum, steel, iron, and other metals that comprise their exteriors are designed to repel lightning. Petty thieves have also taken it upon themselves to capitalize on the Faraday Cage principle, lining their bags, pants, skirts, and other shoplifting mediums with aluminum foil so as to bypass metal and RFID-based anti-theft detectors.

Faraday was more fortunate than most in his field, for his growing contributions to science and technology were not only acknowledged, but rewarded. The year before his construction of the Faraday Cage, Lord William Lamb, 2nd Viscount of

Melbourne, nominated the 44-year-old Faraday to be presented with the government's exclusive Civil List pension. Thenceforth, Faraday was to receive a yearly subsidy of £300 (about £32,000-£33,000 today). Though one could easily argue that this sum, while not insubstantial, is far too scant for a man of his lofty distinction, it was 10 times more than the sum awarded to those on the list today. Even the incomparable poet Lord Alfred Tennyson was awarded only £200.

Lord William Lamb

Later that year, Faraday, who began to dabble in engineering and private consultations, was appointed scientific adviser of the Corporation of Trinity House – chief lighthouse authority of both England and Wales – a position he enjoyed for the next three decades. He was evidently more than competent in this advisory role, for he was soon employed as the lighthouse advisor to the Marine Department of the Board of Trade.

British lighthouses were embarrassingly antiquated in terms of both aesthetics and efficiency, at least compared to other European structures in the 1830s, and Trinity House, though a local landmark of vital importance to the seafaring community, was no different. All their lighthouses featured two fixed beams of lights, which guzzled up costly and wasteful amounts of electricity. It was Faraday who spearheaded a project that aimed to replace Trinity's archaic system with rotating lights, as used by then state-of-the-art French designs. Initially, the special lenses required for the upgrade — invented by French physicist Augustin Fresnel — had to be shipped all the way from France; it was only about 20 years later that they began to source these lenses locally, courtesy of the Chance Brothers, a family-run company in Birmingham.

Unfortunately, like anybody or anything running incessantly at full throttle, Faraday overheated, so to speak. In 1839, at the age of 48, he suffered a crippling mental and physical breakdown, one so severe that he was forced to take a hiatus from his usual work for the next six years. The ailing scientist, whose battered body wrestled with sleepless nights, a poor diet, and countless hours of physically draining and mentally straining work, amongst other factors, found himself unable to stand for extended periods of time, which must have been torturous for such an active spirit. Worse yet, on top of the vertigo spells and throbbing migraines, his memory was beginning to fail him.

Faraday eventually returned to his research in 1845, but he never fully recovered, nor was he the sharp and agile man he once was. Students recalled how a gaunt Faraday shuffled slowly into the room, unable to get through a lecture without reams of detailed notes close at hand. He lamented in his journal, "My [failing] memory both loses recent things and sometimes suggests old ones as new."

This downward spiral was underscored by an event that transpired the following year. In 1846, he published a paper that aimed to explain the electromagnetic properties of light, but due to his inability to support the theory with a clear, mathematical explanation, it failed to attract the attention it deserved. Only when James Clerk Maxwell exhumed Faraday's theory and reinforced it with legitimate equations did the scientific world accept Faraday's theory as fact.

A picture of Faraday holding a type of glass bar he used in 1845 to show magnetism affects light in dielectric material

In spite of his mathematical "illiteracy," Faraday continued to be employed as a consultant for various fields. For example, the Haswell Colliery Coal Mine in Durham County was a questionable firm that found its name slapped onto the local headlines time and time again throughout most of 1844. All the controversy surrounding a strike — which involved, namely, accusations of corruption, dangerous shortcuts, unjust wages, hired thugs known as "candymen," and more — reached a hideous crescendo on the 28th of September. That morning, the mine collapsed, precipitating an explosion that claimed the lives of 95 out of the 99 miners present. Even more tragic was that dozens of them were between the ages of 10 and 13. Faraday was summoned to the terrible scene by Prime Minister Robert Peel and charged with investigating the cause of the explosion, which was later determined to be sparked by flammable coal dust. The entire mine was a death trap, Faraday declared. To give one a better understanding of the negligence, the inexperienced and uneducated miners regularly kept flames next to barrels brimming with gunpowder. The fact that the mine had survived until it did was a miracle in itself.

Faraday urged the government to invest in a proper ventilation system that would vacuum up toxic fumes and flammable dust. Unfortunately, his pleas fell on deaf ears, for coal firms were far more concerned with costs than they were with the safety of their employees. It took another 75 years, as well as another appalling accident – the Senghenydd Colliery Disaster of 1913 – before coal companies began to install ventilation systems into their mines.

In 1848, Faraday was offered the post of President of the Royal Society for the second time, but his Sandemanian faith, which proclaimed it "improper" and "unscriptural" to accept unnecessary riches and accolades, barred him from taking on the position. The same principles also led the pious man to politely decline the honor of knighthood. It took a considerable amount of convincing for Faraday to accept the Hampton Court home that was gifted to him and Sarah by the Crown later that year. Now known to the locals as the "Faraday House," this handsome, three-story brick mansion was where Faraday resided for the rest of his days.

The aging Faraday did some research and experimental work in his spare time, but he dedicated most of his time to fulfilling his supervisory duties. A few years later, Faraday handed to his superiors a proposal that suggested the replacement of existing lanterns with "carbon arc lamps." In 1858, the South Foreland High Lighthouse became the first lighthouse in the world to be powered by electric lamps.

While growingly frail, Faraday used whatever strength he had left to promote scientific changes across the world. A part-time environmentalist, Faraday was often summoned to brainstorm answers for natural hazards and public health crises, such as the Swansea industrial pollution case and the Royal Mint air pollution quandary, to name a few. His most noteworthy achievement in this field was his persistent championing of purifying the River Thames, which had been transformed into a festering, cholera-breeding cesspool of sewage and litter that stunk to the high heavens. In July of

1855, Faraday wrote a letter, "Observations on the Filth of the Thames," to both Parliament and the *Times* newspaper, explicitly detailing his disgust: "The appearance and...smell of the water forced themselves at once on my attention. The whole of the river was an opaque pale brown fluid. In order to test the degree of opacity, I tore up some white cards into pieces, moistened them so as to make them sink easily below the surface, and then dropped some of these pieces into the water...before they sunk an inch below the surface, they were indistinguishable, though the sun shone brightly at the time...I have thought it a duty to record these facts, that they may be brought to the attention of those who exercise power or have responsibility in relation to the condition of our river; there's nothing figurative in the words I have employed...they are the simple truth...If we neglect this subject, we cannot expect to do with impunity; nor ought we to be surprised, if ere many years are over, a hot season give us sad proof of the folly of our carelessness." Thanks to Faraday's letter, as well as pressure from the furious and panicking public, Parliament installed a fresh sewer system, and began a rigorous campaign to restore the Thames to its original glory.

His Life

Born: On the 22nd of September 1791, this person died: August 25 1867 Richmond upon Thames, England (at the age of 75.) Awards and Honors: Subjects of Study: Copley Medal (1838) and Copley Medal (1832)Faraday effect chlorine diamagnetism electromagnetic electrical motor Michael Faraday was an English scientist and physicist. He was born on the 22nd of September 1791 in Newington, Surrey, England He died on the 25th of August 1867 at Hampton Court, Surrey. He was an avid experimenter who contributed significantly to our understanding of electromagnetism.

One of the most famous scientists of the 19th period, Faraday started his scientific career an chemist.

He discovered many new organic compounds, among them the benzene compound, which was the first to make an "permanent" gas, and the author of a manual on practical chemistry that shows the mastery on the scientific aspects in his art.

However his most notable achievements were in the field of magnetism as well as electricity.

He created the very first motor with an electric charge. They also established the link between chemical bonding and electricity as well as the effect of light on magnetism, and identified diamagnetism and gave it the name, the unique behavior of some elements in strong magnetic fields. Also, he was the very first person to create an electric current using an electromagnetic field.

James Clerk Maxwell developed the classical electromagnetic field theory based on his experiments and some of his theoretical theories.

Early Life

Michael Faraday is a native of Newington, Surrey, a rural community that is now part South London.

In 1791 his father, who was blacksmith, relocated from north England to search for work.

His mother was a rural woman with great deal in wisdom, and cool, who helped her son emotionally through the difficult childhood.

Faraday was the fourth of four children, and all of them struggled to get food since their father was often sick and was unable to perform his duties consistently.

Later, Faraday recalled receiving one loaf of bread which was meant to last him for a week.Faraday was nourished spiritually throughout his life by members of the Sandemanians which was a tiny Christian sect.It was a major influence on his perception and understanding of nature, and was the only influential influence on him.

In a Sunday school at the church, Faraday learned to read write, write, and even cipher. He had no formal education.

He started delivering newspapers for a book dealer and bookbinder when he was young, and when he was 14 years old, he was apprenticed to the man.

Faraday took advantage of the opportunity to read some of the books brought in for binding, in contrast to the other apprentices.

He was particularly interested in the article on electricity in the third edition of the Encyclopaedia Britannica.

He constructed a crude electrostatic generator and carried out straightforward experiments with lumber and old bottles.

Additionally, he constructed a weak voltaic pile with which he conducted electrochemistry experiments.

The force that in Physics or natural law slows down motion? What is the opposite of every action? This test in physics doesn't need students to be aware of E = Mc square.

When Faraday was given a ticket to the chemical lecture on the Royal Institution of Great Britain in London and he saw it as an excellent opportunity.

Faraday wentto the lecture, sat and absorbed, jotting down the lectures on his notes, then returned to binding his books in the chance of gaining entry into the science temple.

A request for an interview as well as a bound copy of his notes were given to Davy but the temple did not have an opening.

However, Davy did not forget and provided Faraday an opening after one of his lab assistants was fired due to fighting.

Faraday began his career as a lab assistant for Davy and then learned about chemistry from one of the most skilled scientists of the time.

Davy's greatest achievement was Faraday as per some reports.

His Career

W

When Faraday when he joined Davy at the age of 1812 when and the chemical knowledge of the day was experiencing a major change.

In the years 1770 and 1780, Frenchman Antoine-Laurent Lavoisier was often credited with the invention of modern chemistry, organized the field of chemical research by insisting upon a handful of basic principles.

One of them was oxygen it was a unique element due to the fact that it was the only element which could sustain combustion and was the most fundamental element in all acids.

Davy focused on the decomposition of muriatic (hydrochloric) acid as one of the strongest acids that is known, after the decomposition of the sodium and potassium oxides using a strong current from the galvanic battery in order to find the elements.

Hydrogen and the green gas that helped to support the combustion process and when mixed with water, formed an acid were the products of the breakdown.

Davy reached the conclusion that muriatic acid was devoid of oxygen and was an element and gave its name chlorine.

Therefore acidity wasn't caused due to having an element that produces acid, but more by a different circumstance.

Other than the acid's physical shape is there any other reason that could cause this circumstance be? Davy believed that arrangement certain elements within molecules could be an important factor in determining the chemical properties of the molecules, in contrast to the elements themselves.

He was affected by the atomic theory which would influence the thinking of Faraday in forming the viewpoint.

Ruggero Giuseppe Boscovich proposed this idea at the end of the 1800s and claimed the atoms could be mathematical elements which were enclosed by alternating fields of repellent and attracted forces.A real element was composed from just some of those points as well as chemical elements are composed from many of these points as well as the forces that emerged from them could be extremely complicated.In the process these elements created molecules, and finally, the pattern of force around points created physical properties for compounds and elements.One particular characteristic of these molecules and atoms must be noted: Prior to their "bonds" that held them together broke and they were exposed to significant amounts of tension or tension.Faraday's theories about electricity could be based on these tensions.

The year 1820 was when Faraday's third apprenticeship with Davy was to come to an end.He was as knowledgeable about chemical chemistry as anybody alive at the time point.Additionally Faraday had developed his theories to the point that they could direct him in his studies and were able to master the techniques of laboratory analysis and chemical analyses to the point of total mastery.The scientific community was amazed by the subsequent discoveries.

His early fame was due to his job as an chemist.As due to his fame for being an analytical scientist the chemist was called an expert witness during court instances and also cultivated a clientele who paid for the Royal Institution.He developed the first carbon-chlorine compound, C_2Cl_6 as well as C_2Cl_4, around 1820.In the first substitution reactions caused through "olefiant gas" (ethylene) the chlorine was replaced with hydrogen to create the compounds.Later these reactions would challenge the Jons Jacob Berzelius' main theory of chemical mixture in doubt.)Faraday identified and identified benzene in 1825 in the course of his research into the illuminating gases.He also researched steel alloys during the 1820s which laid the foundation for the development of scientific metallurgy as well as metallography.He created a glass that had extremely high refractive index when working on a proposal for the Royal Society of London to improve the optical quality of glass used in telescopes. The glass led him to the discovery diamagnetism in 1845.He was engaged with Sarah Barnard in 1821, moved into a permanent residence in the Royal Institution, and began an investigation into electric and magnetic fields that could change the nature of physical physics.

The idea that the flow of electricity through a wire produces an electric field surrounding it was revealed in 1820 by Hans Christian rsted in 1820.Andre-Marie Ampere showed that the conductor was enclosed by a cylinder magnetism due to its apparent circular force.Faraday became the very first person to be able to comprehend what a circular force such a nature meant, because there was no evidence of such a force that had been seen before.A magnetic pole must always be in a circular motion around a wire carrying current in the event that it can be isolated.Faraday was able to create an apparatus that confirmed this assertion due to his ingenuity and experience at his laboratory.The the first electrical motor that transformed electric energy to mechanical power.

Faraday began to look into how electricity works in the wake of this discovery.He did not hold the belief the idea that electricity is a fluid that could move through wires the way water moves through pipes, which was not the case for the other scientists of the time.Instead Faraday imagined electricity as a vibration or force that somehow travelled through the conductor due to the result of tensions there.After his discoveries of electromagnetism one of his initial experiments was to run the polarized light rays through a solution going through electrochemical decomposition in order to discover the intermolecular strains that believed to originate from an electric current.He continued to revisit this concept in the 1820s but with no results each time.

Faraday and Charles (later Sir Charles) Wheatstone began working on the theory of sound--another vibrational phenomenon--in the spring of 1831.He was particularly fascinated by the patterns, or Chladni figures, that appeared when a violin bow vibrated iron plates coated in a thin powder.He was convinced that a current-carrying wire was the example of how a dynamic cause could produce a static effect here.The fact that such patterns could be created in one plate by bowing another nearby impressed him even more.His most well-known experiment appears to have been based on this kind of acoustic induction.Faraday wound insulated wire connected to a battery around a thick iron ring on one side on August 29, 1831.He then wound the opposite side using a galvanometer-connected wire.He anticipated that when the battery circuit was closed, a "wave" would occur, which would manifest as a galvanometer deflection in the second circuit.To his delight and satisfaction, he closed the primary circuit and saw the galvanometer needle move.The primary coil had caused a current in the secondary coil.However, he was surprised to see the galvanometer move in the opposite direction when he opened the circuit.In the secondary circuit, turning off the current also produced an induced current that was equal to and opposite to the original current.Faraday proposed the "electrotonic" state of the particles in the wire, which he regarded as a state of tension, in response to this phenomenon.As a result, it appeared as though a current represented either the emergence of such a state of tension or its dissolution.Although he was unable to locate experimental support for the electrotonic state, he never completely abandoned the idea, and the majority of his subsequent work was influenced by it.

Faraday was trying to determine the method by which an induced electric current was created in the autumn of 1831.A powerful electromagnet created by winding the main coil was the focus of his first experiment.He was then able to make permanent magnets to create an current.He realized that a current was created by a coil of wire when an permanent magnet is moved into the and out it.He knew that forces that were surrounded by magnets could be observed spraying iron filings onto cards placed on top of them.Faraday quickly discovered the law that governs the generation of electric currents caused by magnets after observing "lines of force" as lines of tension in the air, medium that surrounds the magnet.The quantity of these lines that were cut by the conductor within the course of a time is the amount of current.By taking leads off the copper disk's center and rim and then rotating it between the poles of a strong magnet, he was able to see instantly that a continuous flow of current might be produced.Since the outside of the disk will be more efficient at cutting lines than inside the circuit that linked the rim and the center would generate an ongoing current.The first dynamo originated from this.It is also considered to be the direct predecessor to electric motors since all that was needed to reverse the direction of the disk and make the disk rotate was to apply electric current to it.

His Inventions

M

Any of Michael Faraday's discoveries or inventions in the field of electrical science or the field of chemistry made him one the greatest scientists of our time.

While Michael Faraday was a modest man who was not looking to become famous, or create his name He was acknowledged for numerous inventions and discoveries. He also made significant contributions to science.

The society we live in today has evolved due to Michael Faraday's innovations and discoveries and his work has served as the foundation of the technology we have today.Many of the bases on that others have built founded on Michael Faraday's contributions to electricity science.He also made major advancements in the field of chemical science.

In recognition of his status and stature, Queen Victoria gave him the knighthood.

In his time He was also asked whether he wanted to be buried at Westminster Abbey, a privilege reserved for queens and kings of the country and famous individuals such as Isaac Newton.But the humble man refused his back.

We should honor Michael Faraday today , for his discoveries and inventions that the man came up with.

Electronic motor: One of Michael Faraday's most important discoveries and inventionsFaraday invented an electric motor for the first time in 1822 even though Ampere and Oerstedt are both believed to have discovered the power of electricity, which creates an electromagnetic field.

Recognized Benzene:

The chemistry of Faraday's early discoveries was a connection.Hexachloroethane, also known as C2Cl6, and tetrachloroethane, also known as C2Cl4, were the first known synthetic compounds made from carbon and chlorine in 1820.After that, in 1825, he produced benzene C6H6, which is an essential component of modern chemistry and the foundation of a great deal of organic chemistry.

Michael Faraday was the one who isolated the an element called benzene. The chemical representation of benzene is his. Electromagnetic induction: Faraday first was the first to make his discovery of electromagnetic induction known in 1831.He created a fundamental transformer that was the subject of an infamous experiment.He put two lengths of wire on the opposite the sides of an iron band in order to make two coils.He connected an electric galvanometer to one of the coils and batteries to the other.He observed the galvanometer kicking when he removed the battery.Transformers are built on this principle and are now referred to as mutual induction.

Generator of magnetic fields: He created an electric generator by putting what he learned about electromagnetism and motors into practice.After studying the lines of force that magnetically connect and the magnetic lines of force, he found that the speed of motion as well as the number of forces cut by the movement were in line with the strength of the current generated by the movement of the magnets.

The Faraday disc, the generator which transformed the energy of mechanical motion into electricity was the first generator that used electromagnetic energy to generate of electricity.

Electrolysis:Faraday discovered his electrolysis two laws through his research and experiments through the combination of his knowledge of chemistry and electrical science.

FIRST LAW According to the Faraday's initial law on electrolysis the amount of the substance that is depositing on the electrode of an electrolytic cell corresponds directly to the amount electrical energy that is passed throughout the cell in the course of electrolysis.

Second Law In accordance with the second law of electrolysis, which was developed by Faraday for a specific quantity of electric power, the quantities of different elements being deposited is related to their chemical weights.

Students who encounter electrolysis during courses will have aware of the laws discovered by Faraday as they gave the basic understanding of electrolysis which provided the foundation for the technology over many years.

Michael Faraday invented and discovered numerous things in his lifetime, based on any standard.In addition, his discovery and innovations were significant because they laid the foundation for many more inventions in the future.Faraday's research served as the base for Maxwell's work as well as his equations, and several other innovations that came later.

It's also fascinating to observe that, as an inspiration source Albert Einstein kept photographs of three prominent scientist on the desk of his, the most famous that was Faraday.

Theory of Electrochemistry

Questions were raised regarding the identities of the various manifestations of electricity that had been studied while Faraday was carrying out these experiments and presenting them to the scientific community.Was the electric "fluid" that appeared to have been released by electric eels and other electric fish, the static electricity generated by a generator, the voltage generated by a battery, and the electromagnetic generated by a new generator all the same?Or were they distinct fluids governed by distinct laws?Even though Faraday was aware that experiment had never adequately demonstrated this identity, he was still persuaded that they were forms of the same force and not fluids at all.Consequently, in 1832, he began what appeared to be a laborious endeavor to demonstrate that all electricities had identical properties and effects.Electrochemical decomposition was the main effect.Static electricity, on the other hand, was a problem. The electromagnetic and voltage electricity were not.Two surprising discoveries were discovered by Faraday while he was investigating the matter further.First It was long believed that chemical molecules dissociate when exposed to the electrical force.Even when electricity was discharged into the air , and did not reach a voltaic cells "pole" or "centre of action," the dissociation of the molecules caused by the movement by electricity in a conductor liquid medium.Second the experiment revealed that the rate that decomposition took place was related to the quantity of electricity passed through the solution.Faraday created a brand new electrochemistry theory based on the results of these findings.He believed that the electric force caused the solution's molecules to be tense, thereby creating an electrotonic state.The motion of particles in the direction of tension, as well as different types of atoms moving in opposing directions, released tension once the force became powerful enough to alter the field of forces which held the molecules and allowed the field to interact with other particles.Therefore chemical affinities of the compounds in solution clearly corresponded with how much electricity passed.These experiments directly led to the two electrochemical laws of Faraday: 1) The quantity of the substance placed on the electrodes in an electrolytic device is related with the amount of electric current that flows across the cell.2) The proportion of the different elements that have chemical equivalent weights what amount of electric power utilized to deposit their respective amounts.

His Later Life And Death

O

One of the most renowned scientist from UK, Michael Faraday, died in 1867, aged 76.That was during the Steam Age was at its height.Numerous cities and villages were connected via rail, factories operated by steam engine steamships travelled across oceans.Faraday's contributions to start of the Electric Age was only beginning.Messages were sent across borders via an electric telegraph.Since the telegraph also linked continents and the first transatlantic cable was installed.In lighthouses electric arc lights were replacing lamps made of paraffin (Faraday was a major part of this) and the first electronic lightbulb was expected soon.Electric generators were in development to meet the world's power requirements as well as electric motors have been utilized to power research vessels and vehicles.

In his final 3 or 4 years of life, Faraday was unaware of the many changes happening that were happening in the outside world.He suffered from mental health issues from the time the age of 18 years old.He was unable to remember even the content of a letter written by an associate in the field after the letter was read since his short-term memory was becoming worse and worse.His problems may have been caused due to mercury poisoning, or any of the kinds of dementia we are all too familiar with today.Mercury was utilized in chemical experiments and instruments which is why it was an integral part of the work of Faraday.

Between between 1820 and 1840, Faraday was the most active researcher on Electrolysis as well as electromagnetic induction.He earned admiration from scientists across the globe as well as the British government as well as Queen Victoria because of the result.While certain people gained wealth due to his research, Faraday took a salary of £2,000 from the Royal Institution of London and did not accept any honors.Mr. Faraday lived a simple life until his demise in his Hampton Court home that Queen Victoria granted him since the cost of buying one for himself.Even although he was an affluent member of the tiny Christian sect called the Sandemanians throughout his entire existence, he resisted the decision to be interred within Westminster Abbey with Isaac Newton. He preferred an exclusive funeral arranged by others belonging to the Sandemanians.

Physics was the area where Faraday was able to make his most important discoveries.He proved his findings in the year 1831. A voltage in one coil of wire could trigger an electric current to another coil.Moving the magnet inside the wire wire could create an electric current like another experiment proved the next year.These findings led to the electricity generator and transformer that are the foundation of the industry.

.

In the Royal Institution, Faraday worked for over fifty years.His duties included public speaking as well as research.In the year 1826, Faraday presented the first of his lectures to children.From the 1840s onwards Faraday was able to make the Christmas Lectures for children a regular celebration. They gained lots of popularity.As Faraday was contemplating the conclusion of his career in science one of the most painful occasions for him was the giving his last lecture at the end of 1861.The Chemical History of a Candle was the most listened-to lectures, was often read and demonstrated his importance to the field of chemistry.

Faraday was not a mathematician, but he did draw diagrams of lines of force to illustrate his theoretical ideas about magnetism and electricity.Nevertheless, many of the physicists and inventors who followed him were influenced by his experimental work and ideas.His works are still remembered at the Royal Institution in London, and whenever we use an electrical device that uses batteries or mains electricity, we should remember him.

Question and Answers About Faraday's Life

T

Have a look through Faraday's story and note of any discoveries the scientist made.

Imagine living in a world without electric power at the touch of the switch.

Examine the significance of Faraday's electromagnetic discovery.

The transformer and generator that was electric weren't patented by Faraday.What did the effect be of this?

Faraday is adored by scientists from all over the world. Despite the fact that Faraday was a member of numerous international scientific organizations, he resisted any domestic honors.What do you think is the cause of this was and what effect do you think it's been on the way we will remember Michael Faraday?

Faraday had only a basic education and did not attend college.How did he come to become so successful in the field of science?

Faraday believed that it is important to deliver talks to young people.Do you agree? What is the thing that attracts children's interest to the lecture?

The quest to unify gravity and electromagnetism, which was also the focus of Einstein's last years , and the minds of many scientists today was the primary principal research of Faraday's final years.

His experiments did not reveal an association, and his research paper that he submitted to the journal was not accepted to be considered for publication.Do you think that carrying out failed experiments is useless?

It is believed that the Faraday Constant as well as The Faraday unit have been named in honor of Faraday.Determine the meaning of these.

Review the information given through Faraday's electrolysis laws.

What is the best way to according to your view, pay tribute to Michael Faraday's death?

The Faraday Electricity Project is helping the world?

E

Electricity is a vital component of modern-day life, which assists us in many ways.

Electricity is utilized for lighting and heating, cooling and refrigeration, in addition to for medical needs as well as the operation of electronics and appliances, computers, as well as public transport systems.We have nothing to do without electricity in our modern times.

How does electricity play a role in our everyday life?

The motion of charged particles, such as electrons (particles that have an electric charge) or protons (particles that carry positive charges) generates electricity that is a form of energy.

For instance, friction creates static electricity.The movement of charged particles happens when a substance is moved towards the opposing direction from one.

Electrons transfer from the inside of your body to the handles on the doors, which results in the stinging fire that we may see when we lift our feet onto the carpet, then rub an electrostatic handle made of metal.

Electricity can be used for a number of things that were previously able to be accomplished using natural gas, coal or the muscles of people, such as making steel and cars, milking cows and more. In the consequence, the total electricity consumption is increasing at an rapid rate.As the consequence, electricity is extensively employed.

Instead of making use of fossil fuels to charge their batteries Electric vehicles

(EVs) utilize electricity.Take online courses in electric vehicles offered by reliable institutions and schools If you're interested in learning the basics of electric cars (EVs) in contrast to cars with internal combustion engine (ICEs).

The significance of electricity in our Daily Lives Electricity is a second source of energy we utilize in our daily lives.

Electricity is created by converting the basic energy sources like coal the natural gas industry, nuclear power solar energy as well as wind power into electricity. This is becoming increasingly important in making life easier for the people and aiding the country's economy recover.

Candles and whale oil lamps to light and cold ice containers to preserve food as well as wood-burning stoves used to heat are all used by humans over the last few decades.

Since the invention of electricity, a variety of daily tasks such as lighting cooling, heating, and running a range of electrical appliances can be completed using electricity.

Scientists' innovations have helped in the creation and development of the use of electricity. Moreover, the discovery of electricity resulted in the development of new devices that changed the technology of the time.

One of the most important inventions in human history is the bulb light invented in 1837 by Thomas Edison.

Samuel Morse created the telegraph in 1837. It was linked to electrical wires throughout India, Europe, and America.

Alexander Graham Bell, a scientist, developed the telephone during 1876 AD. It transmits sound across large distances by turning audio into electrical current, and transferring electrical current via copper wires. Then headphones convert electrical signals into audio signals.

Through the transmission, generation, and utilization of alternating electrical power (AC), Nikola Tesla was also instrumental in reducing costs of transporting electricity over long distances as well as creating electrical devices for homes to operate factory lighting in the interior and to operate industrial machines.

Contemporary agricultural methods are being influenced by electricity and energy as modern electrical machinery is utilized to store and condition grain and grass on farms, and also to process cool and dairy farms.

In order to preserve and storing agricultural crops electric powered equipment has been designed to aid in handling extreme weather conditions that could occur at the time of harvest.

Harvesting grain in days instead months, and drying them using electric fans, the electrical appliances also cut down on the amount of labor required.

Temperature-controlled electric refrigerators are used to store agricultural crops until they are used for longer periods.

The uses of electricity to Modern Life Electricity is a essential element of modern life which helps us in many ways.We are all without electricity in our modern era.We mostly utilize electricity in the following areas:

The home uses of electricity are required to cook and heat water, since they are the main applications, but also entertainment lighting, cleaning, and entertainment.

There are many other electrical appliances used in the home but they are the primary ones.It is utilized for activities such as watching TV heating up the kitchen, washing clothes and bathing, as well as working from home on computers as well as other devices. The use of electricity for residential purposes makes up around 40% of energy use in the world.

The installation of lighting outside security systems, alarm systems, and even traffic lights aid in ensuring safety for the community since electricity is the primary component to guarantee security in major cities as well as other urban centers. Electricity is also a way to create a sense of connection between rural areas from the rest of the.

Medical applications allow you to take pictures of the internal organs inside the body by using X-rays CT scans and MRIs that have decreased the rate of death, as well as the treatment of a variety of ailments by the use of electrical therapy equipment and the use of electrical equipment and other equipment used during surgical procedures.

Agricultural Productivity Electricity helps farmers to increase their productivity through the use of electric machinery. It also allows them to make the most the time they have, increase their production and develop irrigation plans, and boost the quality of their farming.

Electricity helped speed up transport, like rapid electric trains and entertainment options, like television, radio and films. Electricity also facilitated the creation of cutting-edge technology including robots and computers that made life easier for people.

Social Interaction Electricity is useful to connect people living in rural areas that are disconnected from the world. one another and other people in the world. Because they do not have smartphones or other communication device, they are unable to communicate with other cities or aid them in the event of an emergency or if they require help.

Industrial Growth In this time, the usage of electric machines increased and led to an growth in production various items and the capability to run machines across every industry, big and small. Both of these elements have contributed to the expansion of industries as well as improved the social conditions of its members.

The electricity used in the commercial industry covers the cooling, heating and lighting of commercial commercial buildings and squares. Also, it is the power that commercial centers and businesses throughout cities utilize for fax machines, computers printing and copying machines elevators, electrical drawers, in addition to other things. Commercial enterprises also make use of electricity for a range of different purposes.

A Quick Study On Benzene

O

One of the most well-known aromatic compounds is the benzene that has an chemical formula C6H6.

Benzene is an important industrial chemical, which is made by combining oil and coal. It is a natural substance present in many species of animals and plants. It is generated by volcanic eruptions as well as forest fires.

One of the most organic and basic aromatic hydrocarbons, it contains several essential aromatic compounds within its parent compound.The chemical has a distinct smell and is an inert liquid.Polystyrene is the main application to this substance.In addition it being a carcinogen it is believed to be extremely poisonous in nature, and could cause leukemia when exposed to.

The structure of Benzene Benzene has the chemical formula C6H6 which means it is composed by six carbon-atoms, and six hydrogen-H atoms. It also it has an estimated mass of 78.112.The structure contains three double bonds as well as the six-carbon ring appears like the shape of a hexagon.A Corner that's bound to other atoms acts as a symbol for carbon atoms.

Hydrogens are the constituents of the atoms in benzene.This structure is known as having conjugated double bonds since most double bonds within the structure have one bond.Within the hexagonal structure used to symbolize the six electrons the other symbol is a circle.

Benzene is part of the category of hydrocarbons due to it's chemical composition.It is a chemical compound which only has hydrogen and carbon atoms.Benzene is described as a chemical compound made of carbon and hydrogen, with double bonds that alternate that form the circle Its formulas and structure indicate that it is an natural aromatic hydrocarbon.

The background of the chemical benzene Michael Faraday discovered benzene in 1825 while working on an lighting gas.A German chemist by the name of Eilhardt Mitscherlich used lime to heat benzoic acid in 1834. This led to the creation of benzene.A German chemist by the name of A.W. von Hofmann later isolated this the benzene in coal tar.

Chemicals such as benzene are often employed in the manufacturing of common items like detergents, plastics and rubber.

Impacts on Health Benzene is especially detrimental to blood cell-forming tissues.The quantity and duration of exposure to benzene determine the extent to which it impacts the health of humans.

It is known the fact that benzene can be a poisonous substance that could cause immediate and longer-term medical issues.The eyes and the upper respiratory tract are irritated when benzene-vapors are inhaled.

Contact with solvents that contain benzene may cause dryness, itching, split and even cut the skin.